The Open Society Paradox

WHY THE 21st CENTURY CALLS FOR
MORE OPENNESS—NOT LESS

Dennis Bailey

Potomac Books, Inc.
Washington, D.C.

Copyright © 2004 by Potomac Books, Inc.

Published in the United States by Potomac Books, Inc. (formerly Brassey's, Inc.). All rights reserved. No part of this book may be reproduced in any manner whatsoever without written permission from the publisher, except in the case of brief quotations embodied in critical articles and reviews.

Library of Congress Cataloging-in-Publication Data

Bailey, Dennis, 1970–
   The open society paradox : why the twenty-first century calls for more openness-not less / Dennis Bailey.
     p. cm.
   Includes bibliographical references and index.
   ISBN 1-57488-916-8 (alk. paper)
    1. Privacy, Right of—United States. 2. Electronic surveillance—Social aspects—United States. 3. Social control—United States. 4. War on Terrorism, 2001– I. Title.
JC596.2.U5B35   2005
323.44′8′0973—dc22   2004010184
ISBN 1-57488-917-6 (paperback)

Printed in the United States of America on acid-free paper that meets the American National Standards Institute Z39-48 Standard.

Potomac Books, Inc.
22841 Quicksilver Drive
Dulles, Virginia 20166

First Edition

10 9 8 7 6 5 4 3 2 1

To Celeste
for her unwavering support

Contents

Acknowledgments ix

PART I

1 Introduction—The Devil Has a Deal for You! 3
2 Not in My Backyard: The Threat from Terrorism 11
3 Publius Who? Anonymity in an Open Society 26
4 Will the Real John Doe Please Stand Up? A Warning about Identity Theft 39

PART II

5 Your Papers Please: The Case for a Homeland ID 55
6 Smile, You're on Candid Camera: The Case for Surveillance 75
7 There's Gold in Them Thar Data: The Case for Information Analysis 96

PART III

8 Life, Liberty, and the Pursuit of Privacy 117
9 Privacy Lost 130
10 Big Brother Is Watching You 139

| *11* | Invasion of the Data Snatchers | 153 |
| *12* | Information Does Not Kill People; People Kill People | 166 |

PART IV

| *13* | The Open Society of the Twenty-First Century | 181 |

Notes	207
Index	223
About the Author	229

Acknowledgments

Special thanks to those who gave time or had an influence on this effort, including: Paul Rosenzweig, Solveig Singleton, Amitai Etzioni, Sonia Arrison, David Brin, Barbara Doyen, Aaron Plank, Garon Reeves, Randy Cronk, Wayne Reno, Charles Kimble, Ken Graetz, James Thurber, Mitch Frankel, David Coldiron, Brad Graley, Scott Brumfield, Jeff Armstrong, and Kay Bailey.

PART I

The Enemies of Open Society

CHAPTER 1

INTRODUCTION

The Devil Has a Deal for You!

When you're in the outer world, you have to act like them, dress like them, behave like them.

—From an al Qaeda handbook mentioned in the
East Kenyan Embassy bombing trial[1]

In early 2000, when Khalid al-Mihdhar and Nawaf al-Hamzi moved into the Parkwood Apartments in San Diego's Clairmont District in California, they appeared to be two ordinary Muslims trying their best to fit in and stake their claim to the American dream. A local Islamic center helped them settle into the community and afforded them the opportunity to practice their daily religious routines. They enrolled in aviation lessons at Sorbi Flying Club near Montgomery Field, a small San Diego airport, something foreign nationals occasionally did in an attempt to parlay quality U.S. training into well-paying jobs in their home country. They played soccer in the park, ate fast food, had season passes to Seaworld, even frequented a strip club. Essentially, they did nothing that would suggest them to be anything other than two average foreigners trying to make it in the world.

A few things were peculiar about al-Mihdhar and al-Hamzi, however. According to reports, neighbors wondered why they would use their cell phones outside the house whenever they had to make a call. Perhaps it was because they had barely any furniture in their house. With few furnishings, the two would eat their meals on the floor. They'd also play flight simulator games for hours, which neighbors observed through the frequently open front door. Strangest of all, even though the two men apparently couldn't afford to buy furniture, they were often picked up in fancy limousines with tinted windows.

Despite these few quirks, there is little reason to think that two people as seemingly innocuous as al-Mihdhar and al-Hamzi would raise suspicion among Americans accustomed to the idiosyncratic behavior showcased daily in a parade of reality shows and supermarket tabloids. The Federal Bureau of Investigation (FBI), which had a long-time counterterrorism informant in contact with the two men, didn't appear concerned. Apparently the Central Intelligence Agency (CIA) wasn't overly worried either, even after receiving information from Malaysian intelligence that the two had visited with al Qaeda operatives in Kuala Lumpur in January 2000. It wasn't until August of 2001 that the CIA finally raised the alarm and urged the FBI to locate the two men. But by then it was too late. Al-Mihdhar and al-Hamzi had successfully disappeared into American society in preparation for their mission.

On September 11, 2001, when al-Mihdhar and al-Hamzi helped hijack American Airlines Flight 77, the U.S. government and residents of San Diego's Clairmont District finally began to realize that these apparently ordinary men were part of a larger group of al Qaeda militants determined to commit atrocities against the United States. In a post-9/11 world, where Americans are under threat of additional terrorist attacks, the tragic success of al-Mihdhar, al-Hamzi, and the other hijackers forces us to ask a fundamental question: How do we protect ourselves in an open society from those who would use its freedoms against us?

Many who have reflected on 9/11 suggest that the attacks were a watershed event, signaling the changing nature of threats that challenge the United States. During the cold war, knowing who the enemy was and that this enemy was, in fact, oceans away alleviated anxiety somewhat. Today that guarded sense of security has morphed into unease over a faceless enemy that lives in American neighborhoods and is dedicated to jihad against the West. The U.S. military, designed to face more traditional challenges, such as chasing Saddam Hussein out of Baghdad or keeping North Korea on its side of the 38th parallel, appears ill-equipped to counter the shadowy and decentralized enemy in this first war of the twenty-first century. Chapter 2 begins an exploration of this brave new world by looking at modern fanatics who seek to overthrow the West through an ideology that distorts the teachings of Islam.

Overlooked in discussions on the war against terrorism is the distinguishing characteristic that made America vulnerable to the likes of al-

Mihdhar and al-Hamzi in the first place: its openness. We typically think of an open society as the crowning achievement of man. Free markets, human rights, the free exchange of ideas and information, and representative government, these principles of Democracy have brought unparalleled wealth, prosperity, and freedom to U.S. citizens and all but vanquished twentieth-century competitors such as fascism and communism.

Yet 9/11 reminded the world that the openness and freedom valued in Western democracies could be exploited for malevolent purposes. Citizens of Northern Ireland or Israel, who themselves have faced fanatics armed with bombs for decades, know all too well about the susceptibility of open societies to violence.

The United States, the freest and most open of all democracies, must have been a particularly inviting target to a fanatic like Osama bin Laden. With America's porous borders, the ease of mobility and communication, and the relative anonymity enjoyed by U.S. citizens, al Qaeda's leaders knew they could train their terrorist operatives to infiltrate the country and assault any number of targets. Armed with driver's licenses, Social Security numbers (SSNs), bank accounts, and cell phones, the terrorists had the freedom to coordinate the 9/11 attacks knowing that their activities would blur anonymously into the noise of millions of other Americans going about their daily lives. This idea, that terrorists could pervert a cherished value like anonymity into a weapon, prompts the discussion that begins in Chapter 3. The chapter suggests that although a little anonymity is a good thing, untraceable anonymity, particularly the kind developing on the Internet, is giving the likes of copyright violators, spammers, traffickers in child pornography, hackers, and, worst of all, terrorists free reign to roam the digital world. Chapter 4 continues the debate on anonymity by using the growing problem of identity theft to illustrate the risks society faces when individuals are able to shield their identities to escape accountability. The chapter asks the provocative question of whether identity theft is not more of a security, rather than a privacy, issue, and it surmises that a majority of our worst fears regarding personal data and privacy would melt away if identity theft could be eliminated.

The risks inherent to open societies were discussed long before 9/11. The conventional wisdom has always been that a free country like the United States has to allow for a certain amount of lawlessness and violence, knowing that the only other alternative is a police state, one that could effectively stamp out crime, but at the price of our freedom. William Safire of the *New York Times* encapsulates this idea:

When Patty Hearst managed to remain a fugitive for 591 days, that did not mean the FBI was bad at catching fugitives; it meant that America was a free society. In China or the Soviet Union, she would have been captured in days, because it is impossible for ordinary citizens to move about without permission. If our values mean anything at all, they mean that it is better to tolerate the illegal movement of aliens and even criminals than to tolerate the constant surveillance of the free.[2]

The question of finding the right balance between security and freedom can be traced back to the Founding Fathers, who struggled to form a government that would provide for the safety and common defense of Americans without leaving too much power in the hands of the state. *The Federalist Papers* sum up this dilemma succinctly: "In framing a government which is to be administered by men over men, the greatest difficulty lies in this: you must first enable the government to control the governed; and in the next place oblige it to control itself."[3]

Although discussions of how best to create a government that represents the people and protects their natural rights dominate *The Federalist Papers*, the question of security is never far from the minds of the authors. They understood that the informal contract on which society is based provides a guarantee of safety and welfare for the public that is not found in the state of nature. John Locke, the seventeenth-century philosopher who greatly influenced the Founding Fathers, said the following in his "Second Treatise on Civil Government":

> In all states of created beings, capable of laws, where there is no law there is no freedom. For liberty is to be free from the restraint and violence from others; which cannot be where there is no law; and is not, as we are told, a liberty for every man to do what he lists.[4]

The question of ensuring security within a free society has been made relevant once again by recognition of the state of the world following 9/11. The Faustian bargain of trading security for the alluring promise of freedom may have seemed reasonable when the only downside was tolerating the occasional robbery or mugging in a mostly civil society. But in the age where a fanatic might obtain weapons that confer a godlike power of destruction once reserved for superpowers, ignoring the founders' lesson that men cannot be free unless they are first secure may see us on the losing end of a deal with the Devil.

This book argues that technology is creating a world where we may not be forced to choose between freedom and security. New tools of openness that inject transparency into public life are available that can help identify fanatics like al-Mihdhar and al-Hamzi, while allowing average Americans and foreign visitors to go about their lives without interruption.

Part II of this book, starting with Chapter 5, explores recent developments in secure IDs and biometrics, two technologies that can identify terrorists at U.S. borders or prevent them from changing identities if they do manage to enter the country. This chapter bravely argues that the current paper-based system of identification is undermining practically all of the government's efforts in the war against terrorism.

Chapter 6 investigates surveillance technology, such as facial recognition, that can uncover terrorists if they appear at secure sites, such as airports or government buildings. It concludes by asking whether cameras that recognize faces are really any different from small-town communities where most residents are recognized on the street.

Chapter 7 examines information analysis (IA), the third component in an arsenal of openness and a technology that refers to intelligent computer programs that can pick through mundane electronic records to find meaningful clues that terrorists leave behind. Congressional investigators since 9/11 have consistently lamented the failure of intelligence agencies to share information; unfortunately these same investigators have wilted in face of the rhetoric of "Big Brother" and other attacks on programs like Terrorism Information Awareness (TIA) that were intended to solve the information-sharing problem.

Taken together, Chapters 5 through 7 paint a picture of new technologies that, when used together, can remove the camouflage of anonymity that terrorists find in open societies, making them stand out from the crowd in high relief.

Although these technologies are useful in the antiterror campaign, when used appropriately, they can respect civil liberties at the same time. In fact, one surprising idea discussed in this book is that whenever there is an increase in transparency, there may be a corresponding rise in the freedom and mobility that most Americans expect. More openness has the effect of reducing the burdens of security, for example, facilitating the boarding of airplanes for "trusted" American travelers by eliminating the need for an invasive search or intensive interrogation. It's only when authorities have limited information about real threats that they are forced to treat everyone like a suspect or to focus on factors such as ethnicity or religion that have poor validity as predictors of terrorism and create resentment among people. At the same time, transparency ensures that if an airport official mistreats someone or violates that person's rights, the injustice does not go unnoticed.

These technological advances suggest the open society paradox. Although it was openness that left the United States vulnerable to the heinous intentions of radicals on 9/11, ironically it is more openness, especially as it pertains to information, that serves as the country's greatest defense. With technologies that facilitate greater transparency, a would-be terrorist raising money or finding recruits for a planned attack will now have to worry that his source of funds will be electronically traced, his ties to other terrorists uncovered through computer analysis of Internet or cell phone patterns, or his fingerprints or photo linked to an international database of terrorist suspects.

This suggests that the answer to twenty-first-century threats is not to close society by putting more police on the streets, shutting down our borders, or constructing walls of privacy around our citizens; paradoxically, the opening of society casts light into the shadows where terrorists lurk.

The acceptance of more transparency is not a fait accompli for most Americans. If scrutiny can be used to expose terrorists living among us, we must consider that it may also expose aspects of our own lives. Many Americans concerned that greater watchfulness will reveal personal details may need to reconcile this policy of openness with a longstanding desire for privacy.

Part III of this book takes a step in that direction by arguing that a twenty-first-century approach to an open society requires a reconceptualization of privacy. Chapter 8, for instance, challenges the romanticized view that there was once a Garden of Eden of privacy to which we seek to return. Many readers might be surprised to learn that privacy as it is now conceived is a relatively modern development.

Chapter 9 counters the notion that privacy is an absolute right akin to liberty and, therefore, should never be limited, even in matters of national security. The chapter holds that limiting someone's privacy rarely restricts that person's freedom, and when it does, there is usually legal recourse.

Chapter 10 tries to separate fact from fiction by questioning the idea of government as a Big Brother that seeks to destroy our most cherished liberties. In an unexpected defense of the U.S. government, the chapter uses events from the Watergate scandal to the passage of the *USA PATRIOT Act* to argue that the Founding Father's system of checks and balances has served democracy as they intended.

Chapter 11 debunks myths about another privacy bogeyman, corporate America, arguing that restrictions on information would have economic consequences and be an abridgement of a company's free-speech rights. Yet, before the reader concludes that the author has left consumers helpless in face of the prowess of the corporate marketing juggernaut, the chapter takes an unexpected turn, arguing that an opt-in approach applied to all forms of direct marketing would cordon off some personal space for consumers. Just because a corporation has your personal information, it doesn't have the right to harass you with e-mail or calls to your home. As personal information continues to spread, and direct marketing through digital means becomes a global affair, the need to protect consumers from the onslaught will only increase.

Chapter 12 draws the conclusion that in a world flooded with data, we should be less focused on trying to secure personal information and more on making sure it isn't used to restrict our most basic constitutional rights. It is not critical whether Attorney General John Ashcroft knows that I like to travel to Europe or that I prefer to listen to Bach (privacy), but whether he restricts my right to do so in the first place (private choice). It's this conflation of privacy with private choice that causes unnecessary hand-wringing in the media on issues such as identification, surveillance, and IA and that distracts us from the essential question of whether our freedoms are truly being threatened.

Taken as a whole, these chapters suggest that a more limited view of privacy, based on the solid foundation of the Fourth Amendment and paired with a vigorous protection of private choice, will give us the best chance of carving out personal space and safeguarding constitutional freedoms within the realities of a century where "information longs to be free."

Will America be willing to accommodate greater openness? Notwithstanding the barrage of media stories and the harangues by civil liberty groups decrying a loss of privacy, a great many Americans have learned to accept, and in many cases embrace, the increasing amount of transparency found in many aspects of their lives. Twenty-four-hour news networks and their relentless armies of reporters swarm around every scent of a story. A proliferation of surveillance cameras, continually decreasing in size, dots the urban landscape. Several C-SPAN channels cover every move elected officials make. Reality television draws millions of viewers into the private lives of ordinary Americans.

These days, there may be more readiness to hand over personal data and reveal public actions than at any time in recent history. Consumers numbering in the millions use grocery-shopping cards to receive discounts on purchases. A growing number of working parents monitor daycare centers from the office. Online shoppers willingly hand over credit card numbers to complete Internet transactions at the click of a mouse. In many ways, the public has come to acknowledge what the courts have ruled for years: There is little reasonable expectation of privacy in our public lives.

Nevertheless, many skeptical Americans know that an apparatus of openness in the hands of the government creates great temptation for abuse. They have a right to be concerned after witnessing the many perversions of power in the last century. They've watched as the FBI spied on Americans and infiltrated activist groups. They've seen a total disregard for authority by a president during the Watergate scandal. They've read countless history books documenting the scandals, corruption, and threats to liberty experienced in this country since its inception.

Chapter 13 suggests that if the public is to embrace greater openness, it will expect the same of its leaders. Unfortunately, while the U.S. government is pushing for enhanced powers to battle foes like al Qaeda, its members are resisting the implementation of transparency measures that would expose their own actions. The Bush administration has been particularly culpable in cultivating a culture of secrecy through such actions as resisting *Freedom of Information Act* (FOIA) requests and planning enhancements to the *USA PATRIOT Act* without consulting Congress. An asymmetrical transparency, which allows the government to watch over the people and not vice versa, is a historical recipe for disaster.

It may be that there are forces at work driving us toward more transparency that even the U.S. government and other leaders around the world will not be able to resist. The drive for efficiency and the thirst for information in modern societies, undergirded by the exponential growth in technology, will leave little cover for leaders who misuse their authority behind closed doors. For most Americans, a revitalization of trust and accountability will restore the principle of reputation that was once the basis for relations in the village, making it possible to improve collective judgments of who can be expected to act in accordance with societal obligations. And as for the al-Mihdhars and al-Hamzis of the world who seek our destruction? They will be affected most by the discovery that a career in terrorism has become a dead-end job.

CHAPTER 2

NOT IN MY BACKYARD

The Threat from Terrorism

It became clear to many people that after the 9/11 attacks, the United States was in a changed world, one with a new kind of threat and different brand of enemy. During the cold war, America had a clearly defined adversary in the Soviet Union and an unambiguous understanding of that enemy's capabilities and intentions. Moreover, while its Communist counterpart had a nuclear arsenal that could destroy the United States many times over, the U.S. nuclear stockpile served as a deterrent in a strategy of mutually assured destruction (MAD). Although there were a few anxious moments over the years, such as the Cuban Missile Crisis, the public could generally rest soundly behind the protective walls of two large oceans and indomitable American military might. After the fall of the Berlin Wall and the collapse of the Soviet Union, it looked like there would be little to challenge America's security and supremacy in the next century.

The hijackings on 9/11 shattered America's cold war comfort. The terrorist group known as al Qaeda changed the rules of engagement, and by doing so, they brazenly exposed the vulnerability of the world's only superpower. Instead of boldly confronting the United States head on and in the open, the terrorists slinked across American borders and disappeared into the crowds. While America, with a defense budget that exceeds those of the next largest eleven countries combined, had its guns pointed outward, al Qaeda hid behind its enemy's lines and waited for orders to attack. The terrorists discovered that they could take America's greatest strength, its freedom and openness, and turn it into the country's greatest weakness.

There were clear warnings prior to 9/11 that Americans faced a new kind of threat. In 1996 the Air Force, in a report entitled "The Air Force 2025 Project," envisioned a "large-scale terrorist incident on American soil" that would spark "civil demands that authorities use all appropriate national instruments to deter and prevent terrorist acts within U.S. borders."[1] The 2000 National Intelligence Council report, "Global Trends 2015," noted,

> Asymmetric approaches—whether undertaken by states or non-state actors—will become the dominant characteristic of most threats to the U.S. homeland. In 2001, the U.S. Commission for National Security in the 21st Century reported that: "Direct attack against American citizens on American soil is likely over the next quarter century."[2]

Attacks against Americans around the world should have signaled that something deadly might be in the offing. In hindsight, experts should have seen the gathering terrorist storm cloud that included the bombing of the World Trade Center in 1993, the murder of employees outside of CIA headquarters in 1993, the Khobar Towers bombing in Saudi Arabia in 1996, the explosions at two U.S. embassies in East Africa in 1998, the millennium plot to bomb Los Angeles International Airport in 1999, and the USS *Cole* bombing in 2000.

The Face of the Threat

The threat of terrorism is not a recent phenomenon. The twentieth century is replete with examples of terrorists unafraid to commit atrocities against innocent civilians, often as a part of movements to overthrow political institutions or colonial rulers. The century was witness to examples like the Irguns who forced Britain out of Israel, the National Liberation Front (FLN) who drove the French out of Algiers, and the Tamil Tigers who have battled the Sri Lankan government to a stalemate since the 1970s.

A defining characteristic of these and other violent organizations was that they understood the need to win a certain amount of public support for their cause and to act within existing political structures. According to Paul Bremer, who headed the National Commission on Terrorism after 9/11 and was the U.S. administrator of the reconstruction of Iraq, this pragmatic approach necessitated limits on the amount of violence and destruction they were willing to inflict.[3]

The picture has changed with today's breed of terrorist whose primary wellspring of motivation is militant Islam. Of the nineteen foreign terrorist groups on a list published by the State Department, ten are Islamic organizations.[4]

Samuel Huntington, in his classic book *The Clash of Civilizations*, has written about the brewing conflict between the West and Islam. He refers to an Islamic resurgence and the disdain that Muslims have for the West, which they see as "materialistic, corrupt, decadent and immoral."

> The underlying problem for the West is not Islamic fundamentalism. It is Islam, a different civilization whose people are convinced of the superiority of their culture and are obsessed with the inferiority of their power. The problem for Islam is not the CIA or the U.S. Department of Defense. It is the West, a different civilization whose people are convinced of the universality of their culture and believe that their superior, if declining, power imposes on them the obligation to extend that culture throughout the world. These are the basic ingredients that fuel conflict between Islam and the West.[5]

The deleterious notion that Islam is fundamentally incompatible with ideologies of the West precludes a search for common ground and political solutions by its adherents. Instead, it seeds the idea among the minds of young Islamic extremists that the answer to the conflict requires the annihilation of Western civilization. This is evident from writing in a captured al Qaeda training manual: "The confrontation that we are calling for with the apostate regimes does not know Socratic debates . . . Platonic ideals . . . nor Aristotelian diplomacy. But it knows the dialogue of bullets, the ideals of assassination, bombing, and destruction, and the diplomacy of the cannon and machine-gun."[6] As Bremer has said, "These men do not seek a seat at the table; they want to overturn the table and kill everybody at it."[7]

As an indication of the pervasiveness of this attitude in the Arab world, Bernard Lewis, Princeton University expert on Middle Eastern studies, claimed that one year after 9/11, the figure who most embodies this idea is Osama bin Laden: "[He] remains an enormously popular figure not only with the extremists and radicals who form his main support group, but in much wider circles in the Muslim and more particularly in the Arab world," and he is becoming a "Middle Eastern Robin Hood."[8]

Osama has become a hero and role model for Islamic youth. Reports indicated that Osama bin Laden dolls were the most popular toy during the 2002 holiday season in Middle Eastern countries.

Terrorist Tactics

Modern terrorist groups like al Qaeda are so dangerous because they have employed new methods of operation and organization to overcome their military inferiority against the West. Rand analysts John Arquilla and David Ronfeldt have termed the battle against modern terrorist groups a *netwar*.[9]

Many years before 9/11, Arquilla and Ronfeldt were describing the changing nature of the threat faced by the United States. Whereas threats from the past often arose from slow moving and hierarchically organized nation states, the new challenge is from small, dispersed terrorist groups organized in a network-centric fashion without a clear chain of command. Analogous in some ways to a multinational corporation, as many authors have pointed out, al Qaeda differs in that the "branch offices" or "franchises" are able to operate independently of the central office. Although the components of this global conglomerate may be dispersed geographically across wide areas, they stay connected by taking advantage of the latest in communications technology, including cell phones, faxes, e-mail, computer conferencing, and encryption.

One of the difficulties in confronting a decentralized organization is that a network is effective at absorbing a blow against its infrastructure. Because there is little or no centralized command, it is impossible to destroy the network by decapitating its leadership. In addition, if one part is destroyed, other parts continue to function without interruption. A comparison can be made to the Internet, where, when a router goes down causing congestion in one part of the network, traffic is rerouted to other parts. In describing netwar, Arquilla and Ronfeldt have taken the Internet analogy one step further to include a comparison to peer-to-peer (P2P) networks, where users swap music and video files. As the recording industry has seen, these networks are amazingly resistant to efforts to shut them down. Arquilla and Ronfeldt write,

> In some ways, al Qaeda is to terrorism as Napster is to file-sharing. True, declawing Napster did little to put an end to the swapping of MP3 files; smaller, even more decentralized P2P networks have popped up in its

place. Taking out bin Laden could splinter al Qaeda into similar networks—the Gnutellas of terror.[10]

The idea of networked forms of organization applies to more than just terrorist groups. Howard Rheingold has observed the pattern of decentralized organization and swarming in a phenomenon he calls the smart mob.[11] He describes technologies such as text messaging, cell phones, and wireless networks, which allow groups to spontaneously organize and coordinate their movements in unpredictable ways. He offers examples as diverse as Japanese teens using wireless technology to stay in touch during the day despite being in different locations, protesters in Manilla overthrowing President Joseph Estrada in 2001 by organizing demonstrations through text messaging, and activists in Seattle in 1999 during the World Trade Organization summit using communications to avoid routes targeted with tear gas.

The other challenge in confronting a decentralized organization like al Qaeda is that individual members can be notoriously difficult to identify and root out. During the U.S. war in Afghanistan, thousands of al Qaeda members scurried across the Pakistan border and dispersed among overcrowded villages and cities. For every case of success, such as the capture of Khalid Sheik Mohammed, one of the masterminds of 9/11, thousands of others who have passed through al Qaeda terrorist camps over the years remain at large. Some have estimated that at least seventy thousand individuals received training in these camps before the facilities were destroyed.[12] The fact that $25 million and the world's greatest manhunt have yet to turn up bin Laden bodes ill for finding the rest of the al Qaeda apparatchik.

The nodes of the terrorist network reach beyond the Middle East to U.S. shores. Since 9/11, federal authorities have garnered convictions of al Qaeda operatives hidden anonymously in cities like Buffalo, Portland, and Detroit. Estimates given in classified briefings to policymakers put the number of al Qaeda terrorists in the United States at five thousand or more.[13]

Even before the intelligence community redoubled its domestic efforts in the antiterror campaign, there were documented cases of fanatics assimilating into U.S. society. The Muslims of America, a nonprofit group in the United States, was shown to be a front for Jamaat al-Fuqra, a terrorist group committed to waging a holy war against America. Founded by Sheik Gilani, the California organization had several hundred followers living on a thousand-acre plot in the Sierra Mountain foothills. *Wall Street Journal* reporter Daniel Pearl was on the way to meet Sheik Gilani in Pakistan when he was kidnapped and later killed.

NBC Technologies of the Twentieth Century

In light of the sheer audacity of hijacking a modern machine like a Boeing 757 and turning it into a guided missile, one may wonder whether this was a measured action designed to inflict a targeted number of casualties or whether al Qaeda would have preferred to strike an even deadlier blow against America but didn't have the capacity to do so. In other words, would the group consider using something like nuclear, biological, or chemical (NBC) weapons to bring hell to some corner of the world if it could get its hands on just such an instrument of death? Al Qaeda spokesman Suleiman Abu Gheith, quoted in the lawsuit filed by victims of the 9/11 tragedy, suggests that the answer is yes:

> We have not reached parity with them yet. We have the right to kill four million Americans—two million of them children—and to exile twice as many and wound and cripple hundreds of thousands. Furthermore, it is our right to fight them with chemical and biological weapons. America is kept at bay by blood alone.[14]

A number of developments in the world make it more likely that in the near future a crazed fanatic might gain access to NBC weapons and try to fulfill these diabolical desires. For one, the rapid spread of communications technologies like wireless networks and online databases increases the availability of NBC-relevant research found in academic institutions, private companies, and government labs. As manufacturing technology improves, it becomes cheaper and easier to furnish research facilities and to build the individual components needed for NBC weapons. In addition, as the global economy carries prosperity further around the globe, countries like Iran find themselves with the resources to leverage NBC technology, putting more strains on international efforts to prevent proliferation.

Another concern is the long festering issue of already existing stockpiles of NBC weapons scattered across the planet. Although the breakup of the Soviet Union is many years behind us, the fallout of this event still poses concerns as poorer satellite countries continue to bear some of the financial burden of protecting dangerous arsenals of weapons inherited as part of the cold war legacy. There is also the ongoing problem of unemployed scientists looking to shop their expertise to other countries around the world in order to feed their families.

The last decade has witnessed the difficulties of managing stores of developed NBC weapons and their component parts. According to the International Atomic Energy Agency, about four hundred cases of trafficking in radioactive materials have been recorded since 1993, and authorities around the world have seized twenty-six pounds of enriched uranium and nearly a pound of plutonium.[15]

Although the idea that a terrorist group or rogue nation could get its hands on a nuclear weapon is a frightening one, the more realistic scenario, at least in the immediate future, involves the use of a dirty bomb. A dirty bomb is the pairing of conventional weapons such as TNT with a small amount of spent nuclear fuel. The level of expertise needed to manufacture such a weapon is significantly lower than that for a nuclear weapon, making it ideal for a terrorist group.

In fact, reports from British intelligence sources that infiltrated al Qaeda indicate that bin Laden had acquired radioactive isotopes from the Taliban and that scientists affiliated with the group were using a nuclear lab in Heart, Afghanistan, for research.[16] This intelligence was further supported by captured al Qaeda manuals, which showed in detail how to carry out a dirty-bomb attack.

It's likely that a dirty bomb would be constructed out of Cesium 137, which has a half-life of thirty years. If such a device were ever detonated in a city, the results would be devastating. A radioactive substance like Cesium 137 is difficult to contain and easily attaches to soil, concrete, and roofing materials. The Chernobyl catastrophe has hinted at the difficulty of cleaning up such a disaster. According to Fritz Steinhausler from Stanford University, "The Russians tried to clean it up for years, and they eventually gave up. It just wasn't economically viable."[17]

According to one report, a bomb containing ten pounds of TNT and a pea-sized amount of Cesium 137 might shut down forty city blocks for decades.[18] Imagine life in Washington, D.C., if dozens of blocks around the Capitol were closed off for years. Major government departments and agencies in the area would have to shut down and relocate, causing a major disruption across the country.

Although the likelihood of this scenario is hard to judge, it appears that the federal government is taking it seriously. Reports in the media have suggested that the Bush administration has set up a shadow government far from D.C. and has on occasion sent government officials, including Cabinet officials, for overnight stays.[19]

The next danger from the NBC trio is chemical weapons. On March

20, 1995, the Japanese cult Aum Shinrikyo introduced the world to this menace by unleashing the nerve gas Sarin on the Tokyo subway, killing twelve people and hospitalizing five thousand. Indications were that if the group had been successful, their next target would have been New York. Fortunately, the effects were limited because of the group's incompetence in creating an effective delivery device for the gas. During the attack, members used umbrellas to pierce plastic bags containing the gas.

Aum Shinrikyo's members were able to manufacture the chemical agents themselves; today's terrorists, however, may find shopping for them a more appealing alternative. One of the biggest stockpiles of chemical weapons is believed to exist in North Korea. Experts have suggested that North Korea has between twenty-five hundred and five thousand tons of seventeen different varieties of chemical weapons stored within its borders.[20] The fact that North Korea has been caught shipping banned weapons like missiles to rogue nations makes one wonder whether the economically depressed country would resist the entreaties of a terrorist group trying to buy weapons if the price were right.

There are indications that al Qaeda may be trying to develop their own chemical weapons. In 2003, British authorities arrested over a dozen terror suspects who had been manufacturing Ricin out of an apartment in London. Ricin is a deadly toxin that is easily extracted from the castor bean. Many times more deadly than cyanide, an amount the size of a grain of salt is enough to kill an adult. Although Ricin is not easily absorbed into the skin and is difficult to disseminate, the toxin is attractive to terrorists for the ease with which it can be made and stored.

The third category of NBC weapons, and perhaps the most feared by experts, is of the biological variety. Much of the attention paid to biological weapons has focused on anthrax, which, although not contagious, still qualifies as a deadly killer. The attention given to anthrax peaked in 2001 when letters tainted with a strain were sent to the offices of two U.S. senators, a tabloid office in Florida, the *New York Post*, and NBC News. U.S. law enforcement agencies have yet to determine the source of the attacks.

The potential risks from anthrax have been well documented. A 1993 report by the U.S. Office of Technology Assessment (OTA) on weapons of mass destruction estimated that a crop duster carrying one hundred kilograms of anthrax spores could deliver a fatal dose to up to three million people.[21] Experts point out that this would not be a trivial exercise because it would require expertise to dry the spores and adjust their size, as well as to load the anthrax into a sprayer. However, lest we get too comfortable with these technological hurdles, we should keep in mind that knowledge in such matters is rapidly increasing around the world. One report indicated the existence of more than thirteen hundred biotech companies in the United States employing sixty thousand people.[22] Biotechnology has also spread to Europe and to other parts of the globe, and countries like Iran and Syria are eagerly investing in biotech programs.

Perhaps what makes biological weapons such a deadly threat is the ease with which many of them can be spread. Whereas nuclear and chemical weapons tend to stay confined to an area after being dispersed, biological weapons, much like a cold or flu virus, can hitch a ride on a human being and begin an infection spree. In a world with transcontinental flights and high-speed transportation, the mechanism of global contagion is already in place.

One such devilish scenario is described in *The Cobra Event* by Richard Preston, biologist and best-selling author of the nonfiction book *The Hot Zone* about the Ebola virus. The suspense thriller portrays the frightening consequences of an attack on New York City by a lone terrorist using a genetically manufactured "killer" virus. The book was so compelling to former President Bill Clinton that he gave copies to intelligence analysts and even to his political opponent, Newt Gingrich.[23]

Preston brought fears of biological warfare back to the real world of nonfiction with a follow-up work entitled *The Demon in the Freezer: A True Story*.[24] In the book he interviews scientists from the former Soviet Union who provide bone-chilling descriptions of their work in biological weapons programs over the last century. They indicate that the Soviets genetically engineered viruses like smallpox that would resist vaccinations. During the course of their program, the Soviets created enormous amounts of smallpox, enough to infect every human on the planet eleven thousand times each.

Alarmingly, several scientists have admitted to not being able to account for nearly twenty tons of the virus from the Soviet stockpiles. The guess is that it is in the hands of North Korea or possibly a few other countries.

Experts who warn about the dangers of smallpox do so with good reason. Smallpox is nothing short of a nightmarish disease, one that ravages the body without mercy. One writer describes the disease as follows:

> The early symptoms are flu-like: chills, headache, nausea, thirst, bad breath, and constipation. Soon, foul-smelling pustules erupt all over the

body, clustering around the hands, feet and face. Lesions appear on the mucus membranes. Sufferers often become unrecognizable. After about two weeks of agony, roughly a third of patients die, a disproportionate number of them children. Survivors are frequently left blind, deaf, bald, and horribly scarred.[25]

Smallpox can be spread from person to person both through the air and through exchanges of body fluids such as saliva. Because symptoms begin eight to eighteen days after exposure, it is possible that someone could spread the disease without realizing it, making quarantines and other efforts to control the disease of less use.

One bioterrorism exercise carried out in 2001 and entitled "Dark Winter" was developed to simulate the effects of an outbreak of smallpox from a terrorist attack. Organized by the Center for Strategic and International Studies, actors played the role of a news channel named NCN, which documented the outbreak. After watching the scenario, senior members of Congress and the Bush administration were spooked to learn just how quickly smallpox could spread throughout the country, overwhelm government and health providers, and cause mass panic in the streets.[26]

Dark Winter illustrated that in the event of a biological attack, fear among the public could be just as disastrous as the loss of life. In the case of a smallpox epidemic, it remains to be seen whether heath officials could vaccinate the uninfected and quarantine infected citizens before they fled the area of contagion.

The difficulties facing the public heath system became apparent with the outbreak of Severe Acute Respiratory Syndrome (SARS). The virus, which started in Hong Kong in March 2003, was immediately spread to multiple countries by infected travelers. Public heath officials struggled to get the epidemic under control after China hid statistics on the severity of the outbreak and many carriers flouted quarantine rules. One can only hope that small outbreaks such as SARS will help mobilize and prepare the world heath community for a future that might witness the outbreak of a much more devastating and less forgiving disease.

GNR Technologies of the Twenty-First Century

In the high-tech future of the twenty-first century, the world may have more to worry about than the spread of NBC weapons. In April 2000, Bill Joy created quite a stir with an article in *Wired* magazine entitled

"Why the Future Doesn't Need Us," which portended a future where dangerous technologies like genetic engineering, nanotechnology, and robotics (GNR) spiral out of control and unleash devastation on the planet.[27] As one of the visionary architects of the Internet revolution and a cofounder of Sun Microsystems, Joy certainly has the credibility to talk about leading-edge technology. Although some among the high-tech cognoscenti who serve as technological soothsayers dismissed his angst as unnecessarily apocalyptic, Joy has given new light to what inventor and futurist Raymond Kurzweil calls the "promise and peril" of technology.[28]

Joy says his vision of doom was inspired in part by his encounter in the 1980s with the book *Engines of Creation*, in which Eric Drexler lays a conceptual foundation for nanotechnology. The fact that a nanometer is about a hundred thousand times smaller than the diameter of a human hair shows that nanotechnology is about very small sizes, down to the atomic level. The belief is that one day the field will revolutionize manufacturing, allowing products to be built atom by atom and miniature machines to be created at a molecular level.

This atomic precision will dramatically affect fields as diverse as electronics, biotechnology, and energy, and it is no longer the stuff of imagination. Hundreds of start-up companies are working on nanotechnology research, and in 2002 *Science* magazine named the nascent field as its Breakthrough of the Year.

Joy's primary concern is that as GNR technologies like nanotechnology advance, scientists will learn how to create molecular machines that can replicate themselves, such as artificial red blood cells injected into a hemophiliac that recreate themselves as they traverse the body. Much like Prometheus, who stole fire from the gods, machines, through these technologies, will have acquired one of nature's most potent capacities, in this case, the power to recreate themselves. The fear is that like a plague of locusts devouring a field, the potential for uncontrolled reproduction could leave the planet swallowed in what has been described as the "gray goo" problem. Michael Crichton uses this possibility as the basis of his novel *Prey*, where a swarm of microscopic machines breaks out of the lab and replicates madly in the environment.[29]

One can do a thought experiment on the replication dilemma by thinking about today's biotechnology research. It doesn't take much to envisage a scientist engineering a virus for medical purposes, perhaps to carry genes for therapy into the body. Only, instead of a curing a disease, the experiment goes wildly wrong with the virus reproducing quickly, infecting other humans, and resisting vaccination. In fact, some published research shows just how close we might be to this scenario.

When researchers at the Australian National University genetically engineered a mouse pox virus related to smallpox that only affects mice, their goal was to create a contraceptive vaccine that could be used to control mouse populations.[30] Instead of producing the contraceptive affect, the virus became a killer that wiped out all the mice in nine days. More interesting, the virus showed an ability to resist vaccinations. This kind of research is being carried out around the world as scientists look to viruses as vehicles to transport DNA into the body during gene therapy.

When asked about the Australian study, D. A. Henderson, a former U.S. presidential adviser and director of the Center for Civilian Biodefense Studies at Johns Hopkins University in Baltimore, said that instructions on how to modify viruses are regularly published in unclassified journals. In an interview he said, "I can't for the life of me figure out how we are going to deal with this."[31]

It may not be long before scientists can build a virus from the desktop computer in their lab with the point and click of a mouse. In July 2002 scientists demonstrated that one could create a virus if its genetic sequence is known. At Stony Brook University in New York, a team of scientists used bits of DNA purchased from companies around the world to assemble the poliovirus. A *New Scientist* article said, "Dramatic as it sounds, this was no scientific tour de force. All the steps are routinely followed in thousands of labs worldwide. That means anyone armed with the knowledge of a virus's sequence, some science training and a few common tools could recreate the virus in a test tube."[32]

In a nightmare scenario, academics publishing gene segments of viruses could provide the information that terrorists need to create a virus, and private companies selling DNA could supply the materials they need to construct it.

In response to this kind of threat, the United States has acted to restrict research on dozens of substances that could be used in biological weapons. However, as later chapters will discuss, controlling the flow of information in a global network of communications is likely to be a lost cause. Science is a double-edged sword, and for every danger it presents, many more potential benefits are likely to invite additional research from public and private sectors around the world. As scientific knowledge spreads across the globe, keeping information out of the hands of those with nefarious intent will become a Herculean, if not impossible, undertaking.

Realizing the futility of restricting potentially dangerous research, physicist and author Stephen Hawking offers a pessimistic view of the future, predicting that a bioengineered virus will wipe out the human race in this millennium. He said, "The danger is that either by accident or design, we create a virus that destroys us."[33]

Glenn Schweitzer and Carole Dorsch, writing in a 1999 article in *The Futurist*, warned that the availability of powerful weapons means we might need to redefine the meaning of terrorism as we know it: "Technological advances threaten to outdo anything terrorists have done before; superterrorism has the potential to eradicate civilization as we know it."[34]

The Power of One

Whereas in much of the history of warfare, it took the resources of an entire nation-state to wreak havoc on its neighbors, the modern age has made it possible to concentrate the same kind of deadly power in the hands of a single individual at a much lesser cost. As 9/11 has shown, it is possible to inflict mass casualties on a country on a shoestring budget. Some have estimated the cost of carrying out the 9/11 attacks at only $400,000 to $500,000.[35] The return on investment for the terrorists, according to the Federal Reserve, was $33 billion to $36 billion in lost wages and business, property damage, and cleanup in New York City alone.[36]

According to Oxford historian, Niall Ferguson,

> The cheapness and availability of military technology make it easier than ever in history to wage a war. All you need is a handful of violent youths, some small arms and some explosives. In the words of the new National Security Strategy, the new enemy consists of "shadowy networks of individuals [who] bring great chaos and suffering to our shores for less than it costs to purchase a single tank." This is the age of the shoulder-mounted antiaircraft missile and the box cutter that killed 3,000 people: Do-It-Yourself-War, if you will.[37]

It doesn't take much ingenuity to imagine the evolution of the Do-It-Yourself-War. Some dissatisfied radicals from one corner of the world decide to martyr themselves for a fanatical cause. However, instead of strapping explosives around their waist, they decide to inject themselves with a bioengineered smallpox virus obtained from a sympathetic scientist. They could be on a plane to the United States and infecting people before symptoms had appeared and health officials could react. The Defense Department echoed these fears in 1997:

> With advanced technology and a smaller world of porous borders, the ability to unleash mass sickness, death, and destruction today has reached a far greater order of magnitude. A lone madman or nest of fanatics with a bottle of chemicals, a batch of plague inducing bacteria . . . can threaten or kill tens of thousands of people in a single act of malevolence.[38]

Judith Miller, an expert on biological weapons at the *New York Times* and coauthor of *Germs*, claims the cost of germ warfare makes it an inviting option:

> These weapons are cheap . . . cheaper I should say. If you can make a vaccine, you can probably make a biological weapon. . . . But, it is less hard than building the kind of nuclear infrastructure needed to deliver weapons. So, if you wanted to deliver what the strategists call an asymmetric blow to a country like ours, a biological weapon is a very good way to go. Great bang for the buck in biological terms.[39]

Unfortunately, the modern world faces the dilemma that there is little to deter today's terrorist, and there is no sign that they will be in short supply any time soon. As Bill Joy has pointed out, statistically there is no limit to the number of fanatics in the world. Joy quotes William Calvin, a neurobiologist at the University of Washington, who wrote,

> There is a class of people with "delusional disorders" who can remain employed and pretty functional for decades. Even if they are only one percent of the population, that's 20,000 mostly untreated delusional people in the Puget Sound area. Even if only one percent of these has the intelligence or education to intentionally create sustained or widespread harm, it's still a pool of 200 high-performing sociopathic or delusional techies just in the Puget Sound area alone.[40]

Elie Wiesel, a human rights activist who views the threat of terrorism through the spectacles of a Holocaust survivor sums up the modern world's dilemma: "The principal challenge of the twenty-first century is going to be exactly the same as the principal challenge of the twentieth century: How do we deal with fanaticism armed with power?"[41]

Although the spread of democracy has removed most fanatics from leadership positions around the world, new technologies may give individual fanatics a new wellspring of power. Yet, unlike twentieth-century dictators who were firmly rooted in the geography of their nations, terrorists veiled in anonymity operate in a shadowy world that knows no borders.

CHAPTER 3

PUBLIUS WHO? ANONYMITY IN AN OPEN SOCIETY

Anyone who has gone to a Halloween party dressed up in costume knows the scintillating delight of being able to hide one's identity from others. There is a certain delectable pleasure in leaving the baggage of your past behind, avoiding the biases and stereotypes through which others see you, and having the freedom to take on any persona you choose.

Although I'm not the first to make the analogy, it is interesting to consider the ways in which the world has become like a giant masquerade ball. Far removed from the tightly knit social fabric of the village of the past, we've lost the ability to recognize the people we pass on the street. People might as well be wearing masks because we are likely to know very little about them. In other words, these strangers are anonymous to us, anonymous in the sense that not only their names, but their entire identities, are unknown to us—the intimate details of who they are, where they have come from, and how they have lived their lives.

As we'll see, society has found many favorable uses for anonymity, such as in medical testing, where protecting the identity of the patient is usually desired. In a society where many people believe the privacy of the individual is preeminent, one would expect this kind of anonymity to flourish.

However, as we'll discuss throughout this book, anonymity has become one of the central vulnerabilities of an open society. Freedom may have allowed al-Mihdhar and al-Hamzi to rent an apartment, use a cell phone, meet with terrorists overseas, and take flying lessons in preparation for 9/11, but anonymity kept hidden the manner in which these individual actions fit together into a larger mosaic of death.

Starting with this chapter, this book will examine a variety of questions surrounding the issue of anonymity, including whether society can afford to turn a blind eye to its members in an age of terrorism. Furthermore, if we are to consider removing our masks, are there ways to protect our civil liberties? How will we reconcile improved identification with our longstanding conceptions of privacy? And finally, will there be other benefits in addition to security, such as a revival of trust and improved systems of accountability, that might make it all worthwhile?

Anonymity in America

The issue of anonymity is a sensitive one, especially because anonymity has such a long and valued history in America going back to the founding of the country. During the American Revolution, for instance, *The Federalist Papers* were published in New York newspapers under the name "Publius," a pen name for the triumvirate of James Madison, Alexander Hamilton, and John Jay. Thomas Paine wrote "Common Sense" and signed it as "Written by an Englishman." Ben Franklin used many different pen names over the course of his lifetime, including Richard Saunders, author of the *Poor Richard's Almanac.*

Over the years, anonymity has favorably encouraged authors and social critics to express the full force of their views, from Mark Twain, whose real identity was Samuel Langhorne Clemens, to novelist Joe Klein, who as "Anonymous" wrote the novel *Primary Colors.* Klein's novel, a fictional account of a presidential campaign, closely mirrors the Clinton campaign for presidency in 1992.

Anonymity often gives people the courage to engage in behaviors that society seeks to promote. Many police departments offer programs such as Crime Stoppers that allow people to report crimes anonymously. Whistle-blowers in investigations often need assurances of anonymity before turning over information concerning wrongdoing. Anonymous treatment and testing is popular at many medical clinics among patients worried about the negative impact of having their illness revealed.

A hidden identity can shield the victims of crime or disease, allowing them to seek help or advice without the fear of ostracism or retaliation. Researchers rely on anonymity to protect the privacy of their subjects in a study. The media depend on it to protect their sources during investigations.

There are other occasions when anonymity has no clear value other than giving people a little space and a chance to let down their guard. Driving to work among the hundreds of cars on a freeway or disappearing into a crowd on a busy city sidewalk provides a refreshing change from a world that constantly places demands and responsibilities on our individual lives.

The Courts and Anonymity

Although there is no clear right to anonymity described in the U.S. Constitution, the courts have frequently ruled in favor of protecting it in various forms, particularly under the First Amendment's guarantee of free speech. For example, in the case of *Talley v. California* (1960), the Supreme Court struck down a California law prohibiting anonymous leafleting on the grounds that it "might deter perfectly peaceful discussions of public matters of importance."[1] In *McIntyre v. Ohio Elections Commission* (1995), the Court threw out an Ohio statute that prohibited the distribution of anonymous documents designed to persuade voters in an election.[2]

In the post-9/11 case of *Watchtower Bible v. City of Stratton* (2002), the Supreme Court ruled against a Stratton, Ohio, city ordinance that required those going door-to-door to obtain a permit and identify themselves prior to and during canvassing. Not all the justices agreed with this ruling. In dissenting, Chief Justice William Rehnquist said the ruling "renders local governments largely impotent to address the very real safety threat that canvassers pose."[3] Rehnquist cited the 2001 murder of two Dartmouth College professors by teenagers who posed as door-to-door canvassers.

There have been many cases involving the Internet where the courts have bolstered anonymity of speech. In 1997 in *ACLU of Georgia v. Miller*, the Supreme Court rejected a bill by the Georgia Legislature that made certain types of anonymous speech on the Internet a crime. The bill would have penalized people for sending e-mails that falsely identify the sender or use trade names that make it seem that the sender was legally authorized to use them. The Supreme Court ruled that the First Amendment protects anonymous speech and that including one's name is no different from deciding to include other parts of a document's content. According to their ruling, anonymous speech allows people to "avoid social ostracism, to prevent discrimination and harassment, and to protect privacy."[4]

Anonymous online postings have resulted in a number of cases of "John Doe" versus other parties, many of which have sided with the anonymous posters. A decision by a federal court in *Doe v. 2TheMart.com* decided that courts have the power to disclose the identities of anonymous users only if they determine appropriate thresholds have been met. This ruling was made in connection with a stock fraud case where the court found that the identities of individuals who had posted adverse comments about the company on an Internet bulletin board should be protected.

Although several of the cases above illustrate that anonymity has often been upheld when it comes to preserving political expression, a type of speech that the courts believe "occupies the core of the protection afforded by the First Amendment,"[5] there have been cases where a compelling state interest has been found in restricting anonymity. For example, in *Buckley v. Valeo* the Supreme Court ruled that the disclosure of the names of campaign contributors was justified by the government's interest in making sure contribution laws were followed and in helping voters better evaluate candidates based on the sources and uses of funds.[6]

In *Griset v. Fair Political Practices Commission,* the California Supreme Court validated a state statute requiring political candidates to identify themselves on mass mailings. Griset was a candidate for reelection to the Santa Ana City Council who sent to voters on the letterhead of two local associations mass mailings attacking his opponent and failing to identify himself as a candidate.

Although the U.S. Supreme Court refused to hear the *Griset* case, legal scholar Michael Froomkin believes that the more cases like *Griset* and *Buckley* attempt to regulate the political process, the more restrictions on anonymity may be needed to enforce those regulations:

> [I]t seems quite possible that despite the language of McIntyre, the Court would uphold a narrowly tailored statute prohibiting anonymity even in the context of political speech if the statute had clear and palatable objectives. Once down this slippery slope of regulation it is notoriously difficult to find a logical place to stop. A particularly difficult case might be a statute that sought to ban all anonymity in political campaigns on the theory

that if the message is not signed with the actual name of the author, it is impossible to know whether it originated in a political campaign and thus violates campaign finance expenditures limits. . . . Without forcing everyone to sign their messages there may, it could be argued, be no way to monitor what campaigns spend, and thus no way to ensure they do not seek to get an edge by spending beyond the legal limits.[7]

Anonymity, Accountability, and the Internet

Consideration of how anonymity might be used to subvert the political process brings us to a general discussion of the dark side of anonymity, where the veiling of one's identity comes into conflict with a key component of social order, accountability. Unless the members of a society have deep reservoirs of moral fortitude, the temptation to violate the law or abuse power is simply irresistible if one can avoid being identified as the perpetrator. When society loses the inability to hold people responsible for their actions, you have a blueprint for a breakdown in civil order. Adam Smith described this troubling aspect of anonymity many years ago:

> While a man remains in a country village his conduct may be attended to, and he may be obliged to attend to it himself. . . . But as soon as he comes to a great city, he is sunk in obscurity and darkness. His conduct is observed and attended to by nobody, and he is therefore likely to neglect it himself, and to abandon himself to every low profligacy and vice.[8]

In some ways, the Internet is like a microcosm of a city, a virtual world where people can disappear among millions of users and cause mischief without ever worrying about the consequences. In that regard, it serves as a good case study of the incompatibility of anonymity and accountability. For example, while anonymous postings can be helpful to whistleblowers, they can also be used to spread rumors, innuendo, or damaging claims. Stock traders interested in manipulating the price of a stock can easily spread false claims on the Internet about a company.

Everyday millions of people are impacted by one of the most apparent results of online anonymity, spam. Waves of these irritating missives are sent out daily, flooding inboxes and slowing down network resources. Complaining or taking action is difficult for most because spammers resort to a number of tricks to keep their identities hidden. One tactic is to route their e-mail messages through proxy servers, other computers that serve as Internet gateways, so that the source of the message is disguised. Even more sinister are a new wave of viruses that hijack personal computers and use them to crank out e-mail on behalf of spammers.

Online anonymity can embolden the more sinister motives in people. In one case that stirred outrage on the Internet, an anonymous individual posted to the newsgroup sci.astro the supposed last conversations of the crew of the Space Shuttle Challenger before the disastrous explosion in 1986. It turned out to be a fraud. This posting went through one of the earliest anonymous e-mail services, anon.penet.fi., a remailer service run out of Finland. For those readers not familiar with the service, a remailer replaces information in the header of an e-mail to identify the remailer, rather than the e-mail's author, as the sender.

This wasn't the only time this remailer was in the spotlight. The Church of Scientology accused the anon.penet.fi service of enabling a former member who was using the remailer to distribute secret church documents. They pursued the case and eventually obtained a court order forcing the system administrator, Johan Helsingius, to reveal the member's identity. No longer able to promise anonymity to users, Helsingius decided to shut down the system.

One of the most telling cases of anonymity on the Internet involved one of the earliest online communities. The Internet group called the Whole Earth 'Lectronic Link (WELL) started in San Francisco in 1985 and became a favorite among early adopters of the Internet. The group experimented with anonymity and eventually had to abandon it after abusive postings began to proliferate. It's ironic that the technological illiterati, who founded this community and often fight so hard for privacy and anonymity, couldn't make it work among even themselves.

Empowered with anonymity, Internet users can engage in any number of malicious activities with a reduced chance of getting caught. They can spread hate or libel without fear of reprisal. They can break copyright infringement laws through free P2P software networks like Kazaa. Or far worse, they can engage in detestable actions like trafficking in child pornography or electronic stalking.

Untraceable Online Anonymity

As the Helsingius case shows, there may not be a guarantee of online anonymity if a remailer is forced to turn over information that can be traced back to an individual user. For people who only require a low level of anonymity, such as those who use free e-mail accounts on

Yahoo! or Hotmail to send messages under an assumed name (a pseudonym, or "nym" for short), this may not be much of an issue. Nyms can provide anonymity in a number of situations, for instance, for someone with an illness who wishes to receive medical information online.

Traceable anonymity through the use of nyms affords room for accountability by enabling society to uncover a person's identity when necessary. For example, we saw in the *Doe v. 2TheMart.com* case mentioned above that the court established guidelines for when the identity of an online user should be disclosed. In one case, a federal court ruled that the *Digital Millennium Copyright Act* (DMCA) required Verizon to turn over the names of online file traders to the Recording Industry Association of America (RIAA).

It may be that credit cards of the future will use traceable anonymity to allow consumers to purchase online goods without the merchant knowing the individual's identity or credit card number. For example, companies are developing technology that routes digital transactions through a pseudonymity authority, which could verify information on behalf of a merchant without revealing information such as the user's name or address.[9]

Although pseudonyms have the potential to bridge the gap between online anonymity and accountability, there is cause for concern when technology makes it possible to prevent identities from ever being uncovered. A child pornographer covering his tracks on the Internet is an example that comes to mind. However, many would disagree with this view, arguing that untraceable anonymity should be the norm, even when the courts declare that disclosing identity is necessary. In their favor is technology that is beginning to make this desire a reality. For instance, it is possible to link together more than one remailer in a chain so that any one remailer will have information only on the previous link in the chain, not on who originally sent the e-mail. When combined with encryption, this technique vastly improves the chances that anonymity won't be lost if one remailer operator is forced to turn over identity information. Although this is technically feasible, the technique is time-consuming, unwieldy, and requires some computer savvy.

Nevertheless, it's likely that technological developments will simplify techniques like chain remailing and make them available to a broader swathe of the public. Already several software projects are underway that promise to bring easy-to-use, yet powerful, anonymity to the worlds of P2P networks and e-mail messaging, which should be a boon to file traders and spammers, respectively. It may not be long before it becomes possible to download copyright-protected software for free, bombard Internet users with limitless numbers of e-mails, trade in child pornography, or plan a terrorist attack with very little likelihood that one's identity will be discovered. Of all the threats posed by the Internet, the specter of untraceable anonymity is the most ominous.

Crime and Anonymity

The use of anonymity to carry out illicit acts isn't limited to acts perpetrated via a modem and a computer. In the physical world, anonymity is the single most important enabler of criminal activity. David Brin, author of *The Transparent Society*, has made the point that of the number of people committing crimes, only a very few are ever caught, prosecuted, or sent to jail. He reports that out of one hundred felonies, thirty-three are reported, and only for six are arrests made and charges filed.[10] His point is that crime happens not because penalties are too weak, but because criminals can get away with it.

There are many factors that contribute to avoiding culpability for a crime. In some cases, the victim knows who the perpetrator is but doesn't report the person for fear of retribution or a lack of faith in the justice system. In other cases, however, the ability to avoid being identified is a prime factor in enabling unlawful acts to go unpunished. Criminal justice experts have long argued that crime is motivated by desire and opportunity; anonymity drives the latter part of this equation.

Consider someone contemplating robbing a convenience store. The criminal probably expects that when he holds up the store clerk, he is not going to be recognized, even if he isn't wearing a mask. His greatest risk of getting caught is if he is nabbed fleeing the scene or leaves an important clue behind such as a fingerprint. Even with the latter, unless the person has been fingerprinted during a prior arrest, leaving his prints behind may cause little harm. In the end, if the criminal can execute the crime without incident, he can rest assured that he can disappear back into the anonymity of the crowd and avoid the not-so-long arm of the law. It makes one wonder whether crime really does pay!

Richard Ericson and Kevin Haggerty, in their book *Policing the Risk Society*, express the futility that law enforcement faces in a society clamoring for anonymity and privacy:

> One characteristic of public policing is how rarely it deals with crime directly. Public policing has an organized incapacity to do much about crime, and it systematically displaces responsibility for crime control to

other institutions in the name of community policing. This incapacity is related to the institution of privacy, which makes detection difficult, and it helps to explain why, for example, the typical clearance-by-arrest rate for burglary is around 3 percent. If you ask how many people in your city were in possession of marijuana or driving while impaired last night, you are asking a question of the same logical order as "How long is a piece of string?" or "How many grains of wheat are there in a heap?"[11]

Roger Clarke, a noted privacy expert has admitted this problem with anonymity: "Anonymity compromises accountability, in that it undermines society's ability to impose sanctions on miscreants, and therefore reduces the extent to which fear of retribution curbs disapproved behavior."[12]

Terrorism and Anonymity

Documents captured from al Qaeda show that terrorists are well aware that anonymity is a valuable tool for penetrating an open society. For example, an al Qaeda training manual seized by police in England reveals instructions on how to use forged identity records and newly created aliases to move about secretly within a country. It includes suggestions such as shaving one's beard, not adding spouses to identity documents, learning about the profession listed on the documents, and speaking the dialect of the area.[13]

Since 9/11, arrests around the world of al Qaeda suspects reveal the amount of trafficking in false identity documents. For example, Italian officials arrested three men in November 2001 on charges of providing counterfeit documentation and logistics support to al Qaeda terrorists.[14] Ten men rounded up in Morocco in June 2002 were charged with the use of false documents.[15] According to the *Washington Post*, as of July 2002, "more than 380 al Qaeda suspects being interrogated at the U.S. naval base at Guantanamo Bay, Cuba, were captured with counterfeit personal documents."[16]

For the modern-day terrorist, who may be called to travel around the globe to meet contacts and plan attacks, the message about counterfeit identification seems to be "don't leave home without it."

Although many of the 9/11 terrorists used their real names, some used forged documents to obtain a driver's license. Yet, even without taking on assumed identities, they did attempt to cover their trails in anonymity. One approach was to keep their communications secret. Many of the hijackers used Internet access at libraries in the United

States and overseas in order to send messages anonymously. According to one report, a suspect in the 9/11 attacks bragged about the event almost a year in advance to a librarian. "There will be thousands of dead," the suspect said. "You will think of me."[17]

Others in the group used prepaid cell phones and telephone cards or public phone booths to hide the source of their calls. Osama bin Laden used a prepaid satellite phone thinking it would disguise his communications and stopped only after he discovered that the United States was using it to track his movements.

The terrorists of 9/11 were also trained in the intricacies of using encryption to prevent their data from being exposed. The seized laptop of the twentieth hijacker, Zacarias Moussaoui, contained encrypted information that was eventually cracked by the U.S. government.

A report by the Council on Foreign Relations a year after the 9/11 tragedy acknowledged the important role that anonymity and identity crime play in terrorism. One of their key recommendations includes, "Step up efforts to rein in identity fraud by strengthening the anti-counterfeit safeguards in state driver's licenses and passports, passing state laws criminalizing identity theft, and mobilizing 120-day joint local, state, and federal agency task forces to investigate and target phony identification traffickers."[18]

Identity crime also plays a key role in funding terrorist enterprises. According to Dennis Lormel, chief of the FBI's Financial Crimes Section, as quoted in a report by Norman Willox Jr. and Thomas Regan, "Terrorist cells often resort to traditional fraud schemes to fund the terrorists' activities. . . . The ease with which these individuals can obtain false identification or assume the identity of someone else, and then open bank accounts and obtain credit cards, make these attractive ways to generate funds."[19]

Lormel revealed that prior to 9/11 the hijackers had moved over $300,000 through thirty-five different bank accounts that had been opened with fake SSNs.

The Need for Identification

Chapter 1 mentions the assumption held by many that lawlessness must be tolerated because the only other option would be a police state, one that could use identification effectively to halt the spread of crime, but only at the cost of everyone's freedom. In this argument, what David Brin has called the Devil's Dichotomy, people are faced with the choice between security and freedom in a zero-sum game.[20] Rather than looking for solutions that promise both, people are frightened into thinking that shining the light of scrutiny to uncover illegal activity would mean a loss of freedom for average Americans.

Underlying this attitude is the view that if people are forced to identify themselves, the entire foundation of freedom and civil liberties will crumble. The lawyer Alan Dershowitz strongly counters this view by suggesting that people confuse a right to privacy with the concept of anonymity:

> Privacy involves information that an individual should be allowed to keep to him or herself—information such as medical records, sexual preferences and religious views. Anonymity involves something quite different: The right not to be known to the government—in effect, to be able to wear a bag over your head. No such general right is even hinted at in the Constitution, although there is a more limited right to publish anonymously or join organizations without having one's identity disclosed to public view. But what one willingly discloses to public view—one's face, name . . . is protected neither by the right to privacy nor by the limited right to anonymity.[21]

It is certainly true that removing the figurative bag from someone's head can subject that person to abuse and harassment at the hands of those with ill intentions. We saw that the Supreme Court has taken this position in regard to protecting anonymity as a way to enhance free speech. However, the loss of anonymity in no way presumes abuse, and if someone has his or her constitutional rights abridged by being identified, there is often legal recourse. For example, as we discuss in Chapter 12, if an employer discovers the identity of an employee with a medical illness, the *Health Insurance Portability and Accountability Act* (HIPAA) protects the employee from medical-related discrimination.

More important, abuse from the loss of anonymity is an aberration, rather than the norm, as evidenced by the numerous examples of identification that occur in everyday life without incident. Take a second and think about the number of times during a normal week that you are required to present an ID. Taking money out at a bank, applying for a job, entering a place of employment, driving a car, writing a check, claiming Medicare benefits, or visiting a government building are just a few cases in which people must reveal who they are. Identification is so prevalent that many motor vehicle departments issue driver's licenses to people who don't drive. Sociologist Gary Marx suggests that identification underlies the fundamental structure of society:

Thinking of society without personal identities is like a modern building without a foundation. The number of contexts where it is expected and even required far exceeds those where its opposite is required or expected. Indeed failure to identify one's self often leads to suspicion rather than the reverse. As with the Lone Ranger we ask, "who was that masked man?" Just try the simple experiment of wearing a hood or Halloween mask throughout the day and note how it will surface the usually tacit norms regarding identification and a variety of control responses.[22]

Once you are born in the United States, the government grants you an SSN and assigns you an identity, which entitles you to certain privileges through life—to go to public school, claim public assistance, run for office, inherit property, and vote. Only when the identities of individuals are known can the government determine eligibility and then efficiently dole out these benefits in a world overgrown with the vines of bureaucracy.

Identification also enables the government to meet its national security responsibilities. Only by identifying citizens can the government collect taxes to pay for law enforcement and target potential recruits for the military. If a worldwide threat should ever arise that required the United States to reinstate the draft, anonymity would be an obstacle in raising an adequate fighting force.

Jeffrey Rosen, a privacy advocate, has criticized the government for trying to make sure a draft disaster doesn't become a reality. He tells the story of a company, Farell's Ice Cream, which sold a list of people who got free sundaes on their birthdays to a marketing firm.[23] The firm then sold the list to Selective Service, which used it to send registration notices to eligible people. Far from being an invasion of privacy as Rosen suggests, this is a case of the government justifiably trying to find a way to identify citizens eligible for Selective Service. Instead of having an effective way of identifying all eligible U.S. citizens, the U.S. government is put into the unfortunate situation of having to find people based on disparate marketing lists.

Perhaps the most significant justification for identifying citizens is to insure the public's safety and well-being. When people drive cars, fly on planes, care for children, own a gun, or engage in any other number of activities that can affect the lives of others, society has an interest in regulating that person's behavior, including knowing his or her identity. One public comment to the Federal Trade Commission (FTC) stated it well:

> Some in the privacy movement use the word "anonymous" in their verbiage. And there are times when being anonymous is proper. But to go

through life in America as anonymous is not possible. If you want to participate as a citizen, a voter, a borrower of money, a professional, a business owner, a land owner, a driver of motor vehicles, a criminal, a plaintiff or a defendant then the rest of us have a right to know who you are should the need arise, simply because your actions may effect [sic] our lives. That knowledge is essential for the public's general welfare. If you want to be anonymous just do not participate in society. The dangers of an anonymous society greatly outweigh the dangers of an open society.[24]

The need for identification to protect the public safety was evident on 9/11. As of September 2001 the CIA had identified al-Mihdhar and al-Hamzi as terrorists and placed their name on a State Department watch list. Unfortunately, because this information was never communicated to the Federal Airline Administration (FAA), the airline attendants for American Airlines flight #77 were unaware of the true identity of these men, or the fact that these men were two cold-blooded terrorists. The attendants controlled the gate to a deadly modern machine, a Boeing 757, but because of the relative anonymity of al-Mihdhar and al-Hamzi, they let the two pass.

The issue of identification raises a significant question: What responsibilities does being a member of society entail? Although the debate on security and freedom has been dominated by talk about the government's obligation to protect civil liberties while it wages the war against terrorism, it is rare to hear any discussion about the duties that exist for individual Americans. Democracy is a two-sided coin: Every inalienable right it seeks to preserve entails concomitant responsibilities of citizens. Political philosopher John Stuart Mill discusses the matter of the public's duty in his 1859 essay "On Liberty":

> Though society is not founded on a contract, and though no good purpose is answered by inventing a contract in order to deduce social obligations from it, everyone who receives the protection of society owes a return for the benefit, and the fact of living in society renders it indispensable that each should be bound to observe a certain line of conduct towards the rest. . . . These conditions society is justified in enforcing at all costs to those who endeavour to withhold fulfillment.[25]

Is it too much to ask that as a part of this return, society's members confirm that they are deserving of the entitlement? The epidemic of identity theft, discussed in the next chapter, serves to warn us that a society that does not take identification seriously does so at its own risk.

CHAPTER 4

WILL THE REAL JOHN DOE PLEASE STAND UP? A WARNING ABOUT IDENTITY THEFT

Failures in the U.S. system of identification have caused identity theft to rise to the top of the list of the fastest growing crimes in America. Many people are familiar with the horror stories or have known a victim of identity theft. Take this example described at the U.S. Department of Justice's Web site on identity theft:

> [T]he criminal, a convicted felon, not only incurred more than $100,000 of credit card debt, obtained a federal home loan, and bought homes, motorcycles, and handguns in the victim's name, but called his victim to taunt him—saying that he could continue to pose as the victim for as long as he wanted because identity theft was not a federal crime at that time—before filing for bankruptcy, also in the victim's name. While the victim and his wife spent more than four years and more than $15,000 of their own money to restore their credit and reputation, the criminal served a brief sentence for making a false statement to procure a firearm, but made no restitution to his victim for any of the harm he had caused.

Although cases like these motivated the federal government to pass the *Identity Theft and Assumption Deterrence Act* in 1998 in an attempt to crack down on identity theft and make it a federal felony, the overall rate of identity theft has continued to skyrocket. In the United States, identity theft has been described as "the fastest growing crime in the nation" and "the leading form of consumer fraud."[1] According to the FTC, identity theft accounted for 43 percent of all consumer complaints in 2002, and individual losses grew from $160 million in 2001 to $343 million in 2002.[2] Over the last five years, a startling twenty-seven million Americans have fallen victim to identity thieves.[3]

An increasing number of Americans have been impacted by identity theft and are altering their behavior as a result. Recent surveys have found that nearly one third of consumers who have bought products on the Internet have experienced fraud or misuse of credit card information.[4] A National Consumers Union/Dell survey asked, "Why haven't you bought anything online in the last 12 months," and found that 57 percent of respondents were concerned about credit card theft.[5]

Identity theft is a particularly frightening crime because it does not discriminate in the choice of victims. There is no completely effective way for most people to protect themselves. One of my colleagues expressed the hopelessness of the situation by claiming that his only protection was probability. In a lottery of millions of potential victims, he figures the odds are in his favor that he won't be one of the many Americans winning the identity theft ticket.

Even the rich and famous can be caught in the identity theft trap. In fact, because they have more wealth, they may be more likely to be targeted. One story that made headlines showed just how vulnerable society's elite could be to identity theft. A busboy in New York, Abraham Abdallah, employed the Internet and a copy of *Forbes* magazine on "The 400 Richest People in America" to obtain SSNs, home addresses, and birth dates of 217 rich and famous individuals, including Steven Spielberg, Warren Buffett, Martha Stewart, Oprah Winfrey, Ross Perot, and Ted Turner. He used the information to gain access to credit card accounts in an attempt to steal millions of dollars. Authorities tracked him down when a $10 million transfer from a Merrill Lynch account belonging to Thomas Siebel, founder of Siebel Systems, was flagged as suspicious.

According to the Privacy Rights Clearinghouse, which has a Web site devoted to identity theft, the impact on victims can be traumatic. One study found that it took a person fourteen months on average to discover that he or she had been a victim of identity fraud. In many cases, law enforcement is unable to provide much assistance due to the sheer volume of identity fraud cases.

Once victimized, an individual can flag his or her credit reports for fraud, although this may not prevent criminals from running up additional charges. Dealing with credit agencies can be difficult especially when a blemish on your report can start at one of the three major credit bureaus and cascade across hundreds of other smaller credit bureaus. As a result, it takes 175 hours, on average, to clean up a victim's financial records, according to a survey of victims by the Privacy Rights Clearinghouse.

Types of Identity Crimes

A common misconception about identity theft is that it is mainly a case of someone using another person's identity to pilfer money, mostly through stolen credit cards. As we've seen from the FTC numbers above, the financial cost to consumers from this type of crime is extraordinarily high. However, taking on a false identity plays a role in a whole host of illicit behaviors. In his book *The Limits of Privacy*, Amitai Etzioni lists several categories of crime that are related to identity theft.[6] So infrequently are these mentioned as related to identity theft that they bear repeating.

- Criminal fugitives: A report by the inspector general of the Department of Health and Human Services (HHS) highlighted a study by the Federal Advisory Committee on false identification, which revealed that 100 percent of all federal fugitives are associated with identity fraud.[7] According to Etzioni, this includes over half a million criminals each year who avoid trial or incarceration by remaining on the lam. Many of these fugitives commit new crimes while on the run from the law.[8]
- Child abuse and sex offenses: There have been numerous examples of caregivers of children who have been discovered hiding a past of child abuse and sexual offenses. These miscreants take jobs in schools, day care centers, and kindergartens by using false identities to conceal their salacious pasts. Etzioni reports that in 1990, a small sample of states reported that sixty-two hundred individuals convicted of serious crimes in the past were seeking employment as child-care providers.[9]
- Income tax fraud: Tax fraud is an inviting target for criminals who use fake identities to file false returns with the IRS. The cost of

fraudulent tax claims is estimated at between $1 and $5 billion a year.[10]
- Nonpayment of child support: Parents seeking to avoid child-support payments often attempt to move or change jobs in order to avoid being located. To stay ahead of federal databases that store their names, perpetrators assume fake identities that allow them to find gainful employment without discovery. Etzioni reports that 18.6 million cases of families needing help locating absent parents came before the Child Support Enforcement Program in 1994.[11] Not included in this figure are spouses who run off with their children and change their identities to reduce the chance of getting caught.
- Illegal gun sales: Federal law requires that states perform criminal background checks on individuals before they purchase a gun. Felons can use a fake identity to pass this check. A General Accounting Office (GAO) report showed how easy it is for undercover agents to buy firearms using fake IDs.[12]
- Illegal immigration: Etzioni points out between four and eight million illegal immigrants in the United States commit $18 billion a year in benefits fraud through the use of fake IDs.[13]
- Welfare fraud: Cases of welfare abuse include recipients of welfare who use fake identities and report fictitious children to collect welfare payments. Etzioni sites a 1998 GAO report that put the cost of identity fraud in entitlement programs at $10 billion annually.[14]

One category that I would add to Etzioni's list is cybercrime. Cybercrime consists of a range of offenses, including hacking into computer networks, creating and disseminating harmful viruses, stealing intellectual property, and engaging in Internet fraud, each of which is facilitated by anonymity and identity fraud.

Although the word "cybercrime" isn't found in some dictionaries, it has begun to garner a lot of attention from authorities. The total cost of cybercrime in the United States in 2002 was estimated to be $455.8 million, compared to $378 million in 2001 and $265 in 2000.[15] In a 2001 Computer Crime and Security Survey conducted jointly by the Computer Security Institute (CSI) and the FBI, 85 percent of corporations and government agencies had detected computer security breaches within the past twelve months.[16]

One of the fastest growing cybercrimes is Internet fraud. Online auction fraud is the most popular crime in this category, accounting for

46 percent of complaints to the Justice Department's Internet Fraud Complaint Center (IFCC). A popular swindle is to set up an account using another person's name and credit card and then dupe an unsuspecting victim into paying for a product that will never be delivered.

Another devious use of an assumed identity in cybercrime is to con an unwary employee into handing over a password to a computer network. For instance, take the story of Kevin Mitnick, the notorious hacker who spent four and a half years behind bars for breaking into the computer systems of telephone companies. He writes in his book, *The Art of Deception: Controlling the Human Element of Security*, about a technique called social engineering, where hackers contact employees within a company and pose as another person to get access to the system.[17] He claims that people are as easy to hack as computers because their desire to be nice or helpful overrides security procedures.

The federal government has elevated the fight against cybercrime to a much higher priority in recent years. The *Computer Fraud and Abuse Act* (CFAA), which makes hacking, stealing, or destroying data or illegally accessing computers a federal crime, was amended in 1994 to cover any computer connected to the Internet. In May 2000 the FBI and the Department of Justice created the IFCC to provide a clearinghouse for the rapidly increasing number of complaints of Internet fraud.

Many experts fear that the next threat from cybercrime will be cyberattacks launched by terrorist groups. A few days after 9/11, the Nimba virus struck computers connected to the Internet causing billions of dollars in damage to companies and governments around the world. Although the incident was buried in the news of 9/11, it was a clarion call to technology companies like Microsoft, awakening them to vulnerabilities in Internet security.

The Nimba virus and all the other viruses, worms, and plagues on the Internet now too numerous to keep track of are a clear signal to terrorists that a few lines of malicious code launched from virtually anywhere in the world can cost Western countries millions, if not billions, of dollars in economic losses. A site run by the Muslim Hackers Club suggests, "It's time for Muslims on the Web, knowledgeable of hacking, virus making, and all those fringe matters, to join a Club and share their knowledge."[18] The site offers advice to would-be Muslim hackers on how to create viruses, steal codes, and "phreak" networks.

Plans to defend the United States against a cyberattack by terrorists are already underway. A report prepared by the Defense Department found that in the fall of 2001, the FBI began tracking suspicious surveillance of Silicon Valley computers.[19] These were not just any computers, but the machines that electronically control critical infrastructure, such as water plants, electrical grids, nuclear plants, and emergency telephone systems. Experts fear that because they are connected to the Internet, they are vulnerable to hackers, especially as many of these systems were not designed with Internet access in mind. If terrorists took control of a system, they could theoretically shut down water supplies, cut off power to a portion of the country, or turn off the cooling system of a nuclear power plant, triggering a catastrophic meltdown.

Spam: A Variation on an Identification Theme

Spam has been a scourge of the Internet, turning one of the information age's greatest achievements, e-mail, into a headache for many Internet users. In 2003 spam was estimated to cost U.S. corporations almost $9 billion in lost productivity and increased need for greater bandwidth and faster servers to deal with the traffic.[20] Some estimates say that spam accounts for nearly 40 percent of all e-mail that flows across the Internet.

In some ways, the malady of spam has come to resemble the Tragedy of the Commons, a phenomenon first described by Garrett Hardin in 1968.[21] Hardin's tragedy was represented by the story of a community raising cattle on a common pasture. For the individual cattle herder trying to maximize gains, it makes sense to add more cattle to the pasture because he will gain the full benefit of owning each additional animal while sharing the cost of grazing it with the entire community. The tragedy occurs when other rational herders in the community come to the same conclusion and overgraze the pasture, ruining it for all.

Today the Internet resembles an electronic version of the pasture. Although many individuals use the network to send e-mail responsibly, there are those who understand that the cost of adding each additional e-mail to the system is virtually nil as it is shared by the community. As the number of individuals trying to maximize gains increases, the network becomes flooded with spam, causing deleterious results for all Internet users.

Hardin's solution to the Tragedy of the Commons was "mutual coercion, mutually agreed upon," such as taxing farmers who use public land. Although they are appropriate to issues like the sharing of public lands, it seems unlikely that solutions such as taxing will work to combat spam on the Internet. The network has been conceived as a low-cost public service; the taxing of online commerce is resisted and free e-mail through publicly available Internet connections such as libraries is encouraged.

More important, efforts at mutual coercion become infinitely more difficult when one considers the role of anonymity, a factor that Hardin didn't take into account when talking about the Tragedy of the Commons. It may be difficult for a farmer to conceal the fact that his animals are grazing on community lands, but it is no problem for a spammer to mask the source of an e-mail message. Spammers have ingeniously developed all sorts of tricks to cover their tracks, from devising simple "spoofing" techniques that make it appear as though an e-mail comes from a reliable source to hijacking corporate servers to send the e-mail on their behalf. All of these techniques serve to do one thing—undermine the community's ability to hold perpetrators accountable.

Before we can ever deal with spam effectively, we will have to address the issue of identification on the Internet. There have been numerous initiatives to crack down on spammers, including filtering software that attempts to intercept spam before it reaches the inbox, blacklists that go after the Internet Protocol (IP) addresses of spammers, and legislation that seeks to enforce the proper labeling of spam and the inclusion of instructions for opting out. Yet, without fail these efforts fall short because there is no consistently effective way of uncovering the identity of spammers.

Perhaps as spam continues to grow and further infuriate Internet users (and it inevitability will as more people from around the world join the electronic commons), there will be more support for identifying individuals before they log on to the information superhighway. It may be that the technology of digital certificates will assume a more prominent role on the Internet, forcing the operators of mail servers to authenticate their identities and eventually authenticate the individuals who use their service.

Obtaining a Fake ID

Although it is true that masking their identities can serve lawbreakers in crimes as diverse as welfare fraud to cybercrime, the most frightening aspect is how easy it is to pull off. An eighth grader could write a manual on how to pilfer an identity.

The simplest way for an identity thief to set up shop is to steal someone's SSN. Obtaining an SSN is not much of a challenge, considering how frequently they are used in society. One approach to obtaining an

SSN is to go dumpster diving. You might be surprised to know how many pieces of mail you receive that include your SSN. While writing this, I went through my mail and found my SSN in an explanation of benefits report from my health-care provider.

There are any number of ways to obtain someone's SSN. One could steal a person's wallet or purse. An employee could look up someone's SSN in a company's database. How many times do you give your SSN out over the phone to verify that you are the account holder for some account? There is no assurance that the customer representative isn't writing down your SSN on a piece of paper while waiting to access your account.

Technology has proved to be a boon for identity thieves on the prowl for a SSN. Powerful search engines like Google will index Web pages once hidden in the far back-end recesses of a Web site and make them available to the public, unbeknownst to the site's manager. With a few carefully typed search words or phrases, prying eyes can discover lists of SSNs, passwords, or credit card numbers.

Once in hand, an SSN allows a thief to dig up personal information using a credit report. Viewing credit reports isn't difficult; they can be found in many businesses, including auto dealerships, realtors' offices, and banks. Available on the report is a variety of information useful to an identity thief, including an individual's current and previous addresses and account numbers. This information can also be enhanced through the social engineering technique described above. Privacy expert Robert Douglass calls this behavior "pretext calling" and gives the following example:

> Bob Smith is the holder of a bank account at USA Bank. Joe Info Broker obtains from one of dozens of lawful databases, many of which can be found on the Internet, Mr. Smith's full name, social security number, address, and date of birth. Joe Broker then starts calling banks in Mr. Smith's neighborhood posing as someone who has received a check from Mr. Smith. When Joe Broker finds a bank that confirms that Mr. Smith has an account, Joe Broker hangs up. Joe Broker then calls back and identifies himself to the bank as Mr. Smith. The bank, for security reasons, asks for personal information that the bank believes only Mr. Smith would know. Joe Broker armed with Mr. Smith's biographical data is able to convince the bank that he is actually Mr. Smith. The bank then provides Joe Broker with any information he requests on Mr. Smith's account.[22]

In a more modern account of this approach, referred to as "phishing," scammers posing as real companies send e-mail to individuals notifying them of problems with their accounts that can only be resolved if they hand over personal information. The Internet job site Monster.com has battled phishers who post false company information in order to get applicants to hand over personal information like SSNs and addresses. The company has been sending out millions of e-mails to its users, warning them of the scam.

Kinko's discovered another scam that shows the lengths to which personal data thieves will go. The perpetrator installed key logging software on their computers to secretly record the key strokes being typed by customers.

Stealing personal data like an SSN to obtain a fraudulent credit card or to access bank accounts may be harmful to the victim, but it really is child's play in the world of identity crime. The more sophisticated identity crook will want a complete set of documents so that at any time or place he can verify the false identity. One way to do this is to get a birth certificate.

The birth certificate serves as a "breeder document" that allows people to obtain other documents, like a driver's license or passport. A report by HHS's Office of the Inspector General documents how easy it is to receive one. It claims that there are 6,422 different entities issuing birth certificates, and in many states anyone can purchase one without identification.[23]

Many of the entities that issue breeder documents will accept just about anything, including bus passes, employee badges, or library cards. According to the HHS report, some states even have "open" access to public records, allowing someone other than the person on the birth certificate to obtain a copy as long as that person provides the name and birth date found on it. Even, Simson Garfinkel, a privacy advocate usually opposed to central databases, was forced to admit that a lack of standardization and centralization is an issue with identity documents:

> The United States does not operate a central computerized registry of every birth and death in the country. Instead, cities, counties, and states all operate their own record systems. Sometimes records get lost—hospitals burn down, computer files get destroyed. Sometimes there are duplicate records, sometimes there aren't. Many record-keeping systems are antiquated. This lack of centralization can be exploited by people who know how.[24]

Once armed with a document like a birth certificate, thieves can very easily obtain a driver's license. In some cases, motor vehicle departments request only an SSN and proof of residency. Even worse, many states won't even check to see if the SSN is valid, allowing a criminal simply to make one up.

Because the driver's license is used as a de facto ID in the United States, once obtained by a criminal, it opens an unlimited range of possibilities. The criminal can open a bank account, rent a house, buy a car, commit a crime, and engage in many other nefarious activities under an assumed name.

The press has described many shocking accounts of people dragged out of their homes in the middle of the night by police because a fugitive had used their name.[25] In some cases, it can take individuals years to clear their names from state and federal law enforcement databases, and in the meanwhile they face the possibility that any time they use their ID, they may be arrested. In response to the growing problem, California has instituted a program that allows victims of identity fraud to give a PIN number to authorities, who can then validate their case through a toll-free number managed by the state.

Terrorists and Identity Fraud

For terrorists who want to infiltrate an open society like the United States, counterfeit identification can be a very useful commodity. The hijackers of 9/11 knew that they would have to obtain U.S. identity documents if they were going to have the freedom necessary to carry out their attacks. They had to rent apartments. They needed to attend flight school to take flying lessons. They had to open multiple bank accounts to obfuscate their funding. To facilitate all of this activity, the terrorists understood the need to get their hands on the most useful ID in the United States—a driver's license.

The manner in which five of the terrorists obtained driver's licenses from Virginia reveals just how easy it is to get identification. Prior to 9/11, Virginia required that a foreign national provide proof of identity and residence in order to obtain a driver's license. Understanding the weaknesses in Virginia's Department of Motor Vehicles (DMV) system, this group of hijackers led by Hani Hanjour and Khalid al-Mihdhar used falsified information on notarized affidavits to obtain their Virginia driver's licenses and as a result gained the freedom to roam the country as they pleased.

The 9/11 hijackers weren't the first terrorists looking to take advantage of flaws in a Western society's identification system. The story of

Ahmed Ressam provides another case. Ressam was a young man living in Canada who had entered that country with a fake ID as a refugee from Algeria. Canadian security officials, having received word that he was living among a group of terrorist suspects, were about to pounce on Ressam. In the nick of time, Ressam used a stolen, blank baptismal certificate and obtained a Canadian passport using a fictitious name, Benni Norris. He effectively disappeared from the Canadian intelligence radar screen.

With his newfound identity, Ressam was able to leave the country to attend one of bin Laden's terrorist training camps. After his schooling, he reentered Canadian borders with financing from bin Laden and plans for a grand terrorist attack. His mission was to detonate a massive bomb at Los Angeles International Airport during the millennium celebrations.

Ressam's plans went well until he tried to clear U.S. Customs at Port Angeles, Washington. It can only be described as fortuitous that a Canadian border guard noticed that Ressam looked nervous, searched the car, and found the explosives for the bomb in the truck. Experts believe that had Ressam not panicked, his fake passport would have gained him entry into the United States and the chance to carry out his diabolical plan.

Identity Theft as a Security Issue

Many Americans concerned about identity theft have been conditioned to view the crime as an invasion of privacy. Although it is true that someone who accesses your accounts and looks at your personal information is violating your expectation of privacy, the primary impact of identity theft is that your security has been violated.

When someone steals your SSN to access your accounts, it is like they have stolen the keys to your car. If they happen to see some of your personal papers sitting in the passenger seat while driving your car, then a side effect of the offense is that your privacy is violated.

Another analogy is to someone who steals the password to your computer at work to gain access to company files. This is clearly a security violation; privacy may become an issue only if the perpetrator chances upon some of your personal information in the process.

The goal of identity thieves isn't to violate anyone's privacy. That is the furthest thing from their minds. They are simply interested in stealing your identity so that they can pilfer your money or avoid accountability for their actions.

To think about identity theft as a security problem, let's consider how the SSN is typically used. In *The Transparent Society*, David Brin makes the insightful distinction between the use of your SSN as a name and as a password.[26]

In some cases, your SSN is used to find your account in a database. For instance, a bank teller may ask for your SSN to look you up in the system. In this example, your SSN is being used as a name during a process of identification. This is comparable to having a logon for your computer at work. In database jargon, it serves as the "key" that allows the system to find your row of data in the database.

In other cases, your SSN is used as a password during the verification of your identity. For example, when you call the phone company, the customer service representative's computer will automatically look up your file based on the phone number from which you call. However, before the representative will give you any details about the account or allow you to make changes to it, he or she will verify that you are who you say you are, in many cases by asking for your SSN, using it like a combination to a lock.

The SSN was originally designed to be used as a name to track retirement benefits for Americans; however, because the SSN is a quick and convenient way to verify identities, over several decades its use as a password has spread across public and private sectors.

Unfortunately, the SSN was never designed to act like a password. For one, the SSN remains constant. When the system administrator at the office gives you a password to your computer, he or she usually lets you know that it will expire after so many days. If it becomes compromised before it expires, the administrator can create a new one for you. With the SSN, however, you are tied to a single number for your entire life. If a thief gets your number, there is no easy way for it to be reset to something different.

This brings us to the question of how to fight the problem of identity theft. If identity theft can be envisaged as a security, rather than privacy, problem, the need to implement robust security technology to address it as such will meet with greater acceptance. The issue of privacy, although certainly important, distracts people from finding solutions to the essential nature of the problem, which is how to stop someone from stealing the password to a person's identity.

In fact, I believe that if identification were improved in America, identity theft would dry up faster than a creek in the Mohave Desert and, as a consequence, complaints about privacy would slow to a trickle. How many times do you hear about a nosy bureaucrat breaking into a database to look up personal information on a neighbor? Or when was the last time you heard about someone violating the U.S. census database, one of the most off-limits data repositories in the federal government? Rather, as the FTC numbers on consumer complaints demonstrate, most cases involve thieves who are stealing information like credit card numbers that are only useful when paired with fraudulent identification.

Identity fraud is approaching a crisis point in America. As this chapter demonstrates, significant breakdowns in the social order can result when individuals flout identification to avoid accountability for their crimes. Unfortunately, the government's unwillingness to implement comprehensive security measures has allowed the issue to spiral out of control for far too long. Governments and corporations invest millions in securing their networks, databases, and computers from thieves and intruders. Citizens' identities should enjoy the same protections.

As the following chapter demonstrates, new technologies that use biometrics offer promising ways to verify identities securely by using information about the body, such as fingerprints, retina scans, and facial recognition. Biometrics has the potential to replace the SSN as a verifier and to reduce dramatically the number of new identity crimes.

Although biometrics offers great promise for identification, concerns about privacy have limited the uses of many of these technologies. The irony is that as speculative fears about privacy rights slow down attempts to improve identification in America, privacy continues to fall prey to identity thieves.

PART II

Technologies of Openness

After spending the last several chapters detailing some of the evidence for the developing risks of the twenty-first century, we shift our focus to what steps the United States might take in its defense. It's clear that terrorists cloaked in anonymity are a dilemma for a country like America precisely because they take advantage of the freedom that democracy is designed to protect. For some, like William Safire, who believe that it is impossible to go after the terrorists hidden among us without trampling on civil liberties, the only option is to close our eyes, hide behind a wall of privacy, anonymity, and encryption, and hope that fate doesn't deal us a tragic hand.

The other approach, the one that Safire and others most fear, is to restrict freedom and thereby remove it as an advantage for terrorists. Closing society is the immediate reaction of many to an emergency like 9/11, where the instinct is to put more police on the street, shut down our borders, close monuments, restrict access to public buildings, and even seal up our houses with duct tape and plastic sheeting, making it impossible for terrorists to act, but only at the cost of making life insufferable for Americans.

Both shutting our eyes and locking down society carry too great a cost for America to bear; rather, the answer proposed over the next several chapters is that, paradoxically, more openness promises both security and freedom for this current century. The technologies of openness (secure IDs, surveillance, facial recognition, and information analysis) can counter the enemy not by restricting people's freedom and mobility or singling people out because of their race or religion, but by making everyone's public actions more transparent, so that when a sinister plot begins to take shape, it isn't hidden under the cover of darkness but stands out for everyone to see.

CHAPTER 5

YOUR PAPERS PLEASE
The Case for a Homeland ID

If the American public wants to win the war against terrorism, I suggest that eventually we are going to have to confront the issue of identification. Although few have been willing to admit it, the most glaring gap in the U.S. strategy against terrorists is the country's insecure system of identification. For violent radicals who depend on deception and secrecy to hide their every move, a fake ID is a necessary accoutrement, providing a free, all-inclusive pass to a world without accountability.

Having counterfeit identification counteracts virtually every measure that U.S. intelligence can take in the antiterror campaign. The Coast Guard, Customs, and Border Control can have the most advanced surveillance equipment, the most highly trained agents, and the most detailed international terrorist watch list, but with a false identity, a terrorist can stroll across the border without a care in the world.

In fact, this possibility was confirmed by agents from the GAO, who created bogus driver's licenses and birth certificates using off-the-shelf software and applied for credit cards using fake names.[1] Armed with the phony IDs, they posed as U.S. citizens and tried to enter the United States at several different points of entry. Not once did they encounter the Immigration and Naturalization Service (INS) or Customs officials who questioned the authenticity of the documents.

In the analysis of U.S. intelligence failures prior to 9/11, there has been a lot of talk about the need for improved information sharing but very little consideration of the role fake identities play in the battle against terrorists. What good is it if the CIA shares information on a terrorist in the United States if the FBI has that person listed under a different name? It is true that the *USA PATRIOT Act* provided some symbolic attention to identification, for example, by requiring financial institutions to use driver's licenses and passports to identify their customers. Yet, without addressing the systemic failures in identification that allow a teenager, let alone a terrorist, to download and use a fake driver's license off the Internet, efforts against terrorism will continue to be undermined.

In light of the vulnerabilities posed by flaws in America's paper-based, twentieth-century system of identification, it is surprising that so little progress has been made toward implementing improvements. There have been a few efforts in Congress to crack down on identity crime, but because these don't address the underlying problems with identification, they appear doomed to fail.

Regrettably, movement on the issue is stalled because it has become taboo to speak of a national standard for identification. In fact, what has been commonly referred to as a national ID has been so derogated by opponents that the word itself has become a politically incorrect pejorative. Some privacy advocates have "demagogued" the issue to such a point that in many people's minds, the idea of a national ID suggests the authoritarian regimes of Hitler or Stalin. Katie Corrigan, legislative counsel on privacy for the American Civil Liberties Union (ACLU), represents the general approach critics take when discussing the idea of a national ID: "Unlike workers in Nazi Germany, Soviet Russia, Apartheid South Africa and Castro's Cuba, no American faces the demand, 'Papers, please.'"[2]

This chapter puts the hyperbole aside for a moment to propose the creation of a single, secure, standardized ID to replace the patchwork of paper documents currently in existence in America. Call it what you will—a national ID, a secure ID, or a smart card. I'll refer to it as a Homeland ID in the spirit of homeland security and in an attempt to eliminate the visceral response that comes from the use of the term *national ID*. Regardless of how one refers to it, the ID must accomplish one main function if we are ever to shine the light of scrutiny at terrorists: It must securely link individuals to their identities.

However, before we get into a discussion of how improved identification might work, it's important to get up to speed on the technology that promises to revolutionize IDs in this country: biometrics.

Biometrics

Three methods are typically used to authenticate someone's identity. These are usually referred to as (1) something you know, (2) something you have, and (3) something you are.

Something you know refers to a password or PIN number such as the code you use to take money out of an ATM. The drawbacks to passwords are that they can be forgotten, guessed, or inappropriately shared. A 1999 survey revealed that people choose simple passwords that are easy to guess. People were mostly like to select the names of partners, children, or pets (49 percent); 20 percent of men said they used their favorite football team.[3]

Something you have refers to something physical you carry around, such as a smart card or token. Many consultants who travel are given tokens to enable secure access to their laptops. These systems consist of a token ("key") that communicates with a serial-port or USB-based peripheral ("lock") on the computer. Once the user is within a certain range, the key transmits verification information, and the lock enables access to the computer's keyboard and screen.

The third way to authenticate identity is by something you are, which is where biometrics comes in. Biometric technology relies on some personal feature of the body, such as a handprint, fingerprint, facial feature, eye characteristic, or voice pattern. A scanner can capture unique information about a personal feature like a fingerprint, digitize it, and then encode it on a card. This data can then be matched to a scan of the person's feature to verify identity.

For example, John F. Kennedy International Airport was one of the first airports in the country to use iris-scanning technology for employee security. Their scanner captures 247 traits of a person's iris and stores the information on a computer and on the employee's ID card. After swiping their cards, workers stare into a scanner for several seconds until the information from the employee's iris is matched with the information on the card. Once a match is made, the door clicks open. If the scanner fails to make a match, alarms sound.

Another biometric technique involves retina scanning. This is similar to iris scanning, although the procedure is much more invasive. Retinas also have the drawback of changing over time.

Face recognition software works in a similar way. It creates a matrix of numbers, including the distance between eighty or more distinctive points on the face. To achieve a match, a certain number of those distances must correspond with a photograph in the database.

A whole range of other biometric technologies is currently under development. Some researchers are looking at identifying individuals by measuring the color spectrum emitted by their skin. Another method uses body proportions such as leg length and waist width. DNA from body fluid or hair samples is still another approach that can be applied, and new technologies on the way will allow for instant identification, such as at the scene of a crime. For instance, recent breakthroughs will soon allow investigators to extract enough DNA from a fingerprint to uncover the genetic identity of an individual.[4]

Use of biometric technology has been gaining momentum in the United States and around the world. The *Aviation Security Act*, which passed in October 2001, mandates the use of fingerprint biometrics for background checks on airport employees. Educational institutions are using fingerprint scans to allow students to buy lunches and log on to computers. Banks are replacing PINs and bankcards with fingerprint and iris scans that identify customers during transactions. Supermarkets are using the technology to automatically debit a shopper's credit card on checkout or to confirm the identity of customers using checks. Manufacturers are shipping computers with smart card readers and fingerprint scanners that cost less than $100, and Microsoft is building biometric support into its next release of Windows.

Issues of accuracy and reliability must still be overcome if the use of biometrics is to become widespread. For instance, sometimes readings of biometric measurements result in false positives (i.e., confusing individual identities) or false negatives (i.e., failing to identify someone). Of course, in case of error, other methods of identification can be employed, such as in the case of facial recognition when a human can compare a person's face with the image in the database to determine if there really is a match.

There have also been cases of people outwitting the technology, for instance, by using plastic fingerprints to fool a fingerprint system. As the technology improves, however, it is likely to become much more difficult and costly to overcome by forging or other means. For example, some companies are rolling out technology that uses an infrared light source to measure the pattern of veins in a person's hands. Although veins are as unique as a fingerprint, they are much harder to fake. A simple plastic finger isn't going to do the trick for the thief of the future.

Is the public ready to accept biometrics as a means to foil the efforts of identity thieves? A survey designed by privacy advocate Alan Westin and conducted by the U.S. Bureau of Justice Statistics seems to suggest that they are as long as the appropriate safeguards are in place.[5] According to the results, the respondents found the following uses of biometrics acceptable:

- Checking the identity of an individual buying a gun against a database of convicted felons (91 percent)
- Verifying the identity of those making credit card purchases (85 percent)
- Withdrawing funds from an ATM (78 percent)
- Accessing sensitive information, such as medical or financial files (77 percent)
- Conducting background checks (76 percent)
- Screening out those banned from gambling or professional card counters in casinos (56 percent)

Blueprint for a Homeland ID

Are there ways to leverage advances in biometrics to enhance identification without greatly increasing risks to the freedoms and liberties guaranteed by the U.S. Constitution? Although some critics simply refuse to consider how identification might be improved, and a few probably want to abandon IDs altogether, many Americans can likely agree on an upgrade to the current system if the changes are reasonable and the proper safeguards are put in place. An approach for a Homeland ID as envisaged here follows some general guidelines:

1. Maintain a single standard for creating and issuing cards and use state departments of motor vehicles (DMVs) to issue them.
2. Use a lightweight, tamper-proof smart card with a microprocessor.
3. Use a biometric measure like a fingerprint or facial scan to verify a person's identity accurately.
4. Limit information in state DMV databases to only a few items, like a name and a template of the biometric measure.
5. Couple strict oversight with rigorous safeguards to protect privacy and prevent abuse.

Such a program as described above would seek to add to and improve existing identification systems instead of replacing them. Rather than creating a bureaucracy at the federal level to manage a Homeland ID, DMVs at the state level could be in charge of administrating the program, as has been proposed by the American Association of

Motor Vehicle Administrators (AAMVA). Removing control of the program from the federal government and keeping it decentralized among states would have symbolic value by reducing Orwellian fears and giving the public a level of comfort in dealing with a local government office.

A Single Standard for Identification

Of course, having fifty states administer a Homeland ID would require some standardization so that criminals and terrorists could not just move to the state with the poorest security. I discussed in the previous chapter the pitfalls that can arise when thousands of different entities in the United States issue identity documents. If every state had adhered to the same policies prior to 9/11, it may not have been so easy for the hijackers to get identity cards. At the time, Florida required no proof of residency to get a license; thirteen of the hijackers knew this and took advantage of that knowledge.

Rules and guidelines for a Homeland ID would include everything from the card design, encryption technology, and procedures for obtaining a card, to how the database is designed, secured, accessed, and audited.

A Lightweight, Tamperproof Smart Card

A Homeland ID card would be akin to a driver's license on steroids. Most of the public has a driver's license and might be persuaded to upgrade it, much as they might update software on a computer. The new card would build on the successes of currently available smart card technology that includes built in microprocessors to store data.

When a person needed to validate his or her identity, that person would swipe the card through a scanner and have a face, finger, or eye scanned. A scanner might not even be needed. For instance, Germany is rolling out a smart card no thicker than a credit card that can scan fingerprints entirely on its own.[6] The card, with technology developed by Infineon, can read the grooves and ridges of any finger placed on the card's surface.

Advances in cryptography enable information to be encoded on the card, rendering it useless to hackers. When an issuing authority such as the DMV first created a card, it would use a private key to produce a unique, encrypted digital signature. Even if a counterfeiter tried to encode his or her own biometrics on the card, the digital signature would reveal the forgery because it would be impossible to duplicate without the original key.

Once the public becomes comfortable with the Homeland ID, some citizens might want to use the card for different applications. The computer chip would provide enough space that data could be partitioned on the card and protected by the equivalent of minifirewalls, guaranteeing that companies or government agencies could only access the data for which it had rights. For instance, individuals with health conditions could have their medical histories on file for easy access by a physician in the case of an emergency. No one other than hospital staff would have a key to this information.

The Progressive Policy Institute suggests that smart cards could lay the foundation for a digital government initiative, allowing people to interact online with the government with their IDs.[7] Voter registration, hunting and fishing licenses, library cards, benefit information, and government permits are a few of the examples of information that could be maintained on a smart card. Other countries to have taken this approach include Malaysia, whose MyKad card contains encoded personal information and uses public key infrastructure (PKI) for online transactions in both the public and private sector.

A forward-thinking program could protect privacy by implementing pseudonymity into the card, enabling the verification of someone's biometrics without revealing that person's identity. For example, a police officer who pulled someone over for speeding could scan the driver's Homeland ID to see if it raised any flags against the National Driver Registry (a database of those with revoked or suspended driver's licenses) or against terrorist watch lists without ever discovering the driver's identity.

Limiting the Data

There is no good reason to have DMV databases store profiles of the individuals to whom they grant IDs. The DMV should focus on doing one thing well, and that is granting secure identification to those who are eligible. The building of terrorist watch lists or mining databases to uncover a terrorist plot in motion should be left to other government bodies that specialize in these matters.

One of the biggest concerns with identity theft is the ease with which individuals can obtain multiple IDs from different states. Although I agree with the general approach of keeping as little information on people in a government database, the one piece of information that it is necessary to store is a biometric measure. This way, a comparison can be made with other state databases to ascertain whether someone has already received an ID in what is called negative identification. The state of Illinois is following this tact, using a facial recognition system from Viisage that scans a database of thirteen million photos looking for matches. The company claims the system has already prevented thousands of people from obtaining multiple IDs under different names.[8]

One way to secure a state database of biometrics, besides using standard security protocols such as encryption and auditing, is to store the thumbprint or face scan as a template, rather than as a complete digital copy. This would provide enough information to determine if there was a match, but not enough detail to recreate the print if the database's security were violated.

The Initial Implementation of a Homeland ID

A significant investment in time, money, and effort would be required to upgrade the driver's licenses of U.S. citizens to Homeland IDs. Such an effort would include implementing the technology, training DMV employees, educating the public, and passing the legislation to create, oversee, and safeguard the program.

Particularly difficult would be the effort of verifying the identity of individuals and resolving disputes before enrolling them in a new identification program. As the Progressive Policy Institute points out, the focus should be on "verification rather than customer convenience."[9] As we saw with some of the 9/11 hijackers who used false affidavits to obtain identity documents, some people will misrepresent themselves. Although many of these instances might never be resolved, at least the system would be an improvement over what exists today in that people would be permanently tied to a single identity and unable to switch to a new one at a later date.

In most cases, the public might not notice much of a difference with a Homeland ID card. They would use it just like they use a driver's license, except in some cases they would provide a fingerprint, iris, or face scan when showing it. Initially, biometric scanners would be installed at locations requiring the greatest security, such as airports, government buildings, public utilities, gun dealerships, and so on.

Much like a credit card reader that dials into a bank, these machines would connect electronically to a national terrorist watch list. They would also check for revoked or expired licenses, a measure that would have prevented Mohammed Atta from boarding a plane on 9/11 because a warrant was out for his arrest, and his license had been revoked for failure to show up for a court appearance.[10]

Eventually, as the public becomes more comfortable with and even desires the use of biometrics, scanners might start turning up in just about any place of interaction, including retail stores, restaurants, banks, post offices, hospitals, and ATMs.

A final piece of the puzzle would be to combine the Homeland ID with restrictions on the use of SSNs. Even if the public had a secure ID card, many private companies and government agencies might continue to use the SSN as a password to individual accounts, especially for online or phone transactions. This is where proposed legislation in Congress limiting the use of SSNs can complement a Homeland ID program. Congress should force companies to use alternative methods of identification, such as PIN numbers and passwords, at least until a future arrives in which biometric scanners become standard equipment on PCs and exist in everyone's home and office.

Criticisms of a Homeland ID: Ineffective against Terrorism and Crime

If a national identification program is ever to be implemented, proponents will have to answer its vocal critics. For instance, there is what I call the First-Time Terrorism Argument. Many opponents of national IDs have argued that there is a risk of giving identification to an unknown terrorist, thus granting him or her membership in the ID club and the freedom to move throughout society and to bypass security checks.

This argument has the smell of a red herring. Of course, a national ID isn't going to prevent a terrorist, especially one who has never been suspected of anything, from committing a terrorist act. If someone from an al Qaeda sleeper cell is a peaceful member of U.S. society, then one day decides to crash his car into a crowd of people, there is very little, short of hiring a psychic, that can be done to prevent it from happening. Even criminals have to plot and commit their first crimes, and until they do, law enforcement is helpless.

The same argument could be used against surveillance cameras. A video camera in a parking lot may not prevent a crime from happening; that does not render the camera useless. A surveillance system can alert law enforcement to a crime in progress. It may even provide clues to find a kidnapped victim or identify the criminal during an investigation. There's no denying the value of capturing the 9/11 hijackers on video in uncovering their identities and retracing their steps prior to the attack.

Furthermore, no one ever suggested that someone without a long-standing record of solid citizenship in America should be granted the freedom to skip past security checkpoints. There is a big difference between letting someone's grandmother avoid a pat-down at the airport and waving through a foreign visitor who has only been in the United States for a short period of time, as was the case with the 9/11 hijackers. A secure ID will never be a carte blanche for unlimited access to secure sites; instead, it will be a part of an overall plan that includes information analysis (IA) and human intelligence to make improved judgments as to who is and is not a risk.

The Homeland ID can help foil first-time terrorists once they start planning an attack. Whereas today a terrorist can obtain a counterfeit ID and effectively disappear down a trail of false identities, a Homeland ID becomes a virtual ball and chain. Once investigators uncover evidence of a terrorist plot in motion, the targeted suspect, unable to switch identities, will be trapped within U.S. borders. Unable to evade watch lists, it will become much harder for the terrorist to buy a gun, board a plane, or leave the country. Furthermore, IA techniques to locate the individual will become more effective because it will be easier to piece together electronic clues from a single identity.

It is even possible that one day the United States may prevent potential terrorists from joining the ID club in the first place. As more countries begin to use biometric technology and the war against terrorism expands globally, an international watch list of terrorist suspects will grow. For instance, at the 2003 G8 Summit, member nations spearheaded by the United States agreed to work together to implement a common biometric standard in passports.

Taking international identification a step further, Interpol, the international police agency, is working to link all 181 member countries to an Internet-based clearinghouse that will flash suspects' digital fingerprints, pictures, and even DNA profiles by 2004. Intelligence agencies around the world are leaving the Stone Age of telex machines and regular mail and are upgrading to the digital technologies that will make it much harder for terrorists to use international escape routes to avoid the law.

Unfair Immigration Policy

As terrorism has become a global threat, a Western country's borders are its first line of defense. Although international efforts are underway to create global terrorist watch lists that will allow countries to screen their foreign visitors more closely, the trafficking in fake travel documents greatly undermines these efforts. Pakistan may share intelligence information about an al Qaeda suspect with the United States, but if the individual uses a counterfeit paper ID, his name will never match the watch list, and he's likely to slip through the hands of border agents. We've already seen that without much effort, GAO agents were able to cross U.S. borders with fake IDs.

The need for improved border security is an issue with much room for consensus. Even Phyllis Schlafy, a right-wing conservative who speaks out virulently on privacy issues, has recognized the need for improved identification of foreigners visiting the United States.

> It's important for Americans to understand that the 9-11 hijackings are a problem of the U.S. government allowing illegal aliens to roam freely in our country. . . . We should have a computerized database of all aliens entering the United States, whether they are tourists, students, or workers, and a tracking system that flags the file when a visa time expires. Aliens should be required to carry smart ID cards that contain biometric identifiers, the terms of their visas, and a record of their border crossings and travels within our country, similar to the rubber stamps used in all passports.[11]

The government seems to agree with Schlafy. As required by the *Enhanced Border Security and Visa Entry Reform Act* of 2002, the U.S. VISIT program was created to collect information on foreign visitors to the United States, including fingerprints, photos, and eventually iris scans. This program works in conjunction with the Student and Exchange Visitor Information System (SEVIS), which requires educational institutions to record information on foreign students' school activities. Had the system been in place prior to 9/11, it may have led investigators to Hani Hanjour, one of the hijackers, who failed to show up for school as required by his student visa.

Critics have vehemently charged that changes to immigration policy will unfairly target visitors of Arab descent. Although the fact that many of the hijackers were from countries like Saudi Arabia provides fodder for those who argue for a more careful screening of this group, the U.S. government attempts to screen all visitors alike. Many immigrant advocacy groups are behind this effort because they realize that technology like biometrics will enable the United States to avoid singling out a particular group based on race or ethnicity. The paradox is because technologies like biometrics can improve tracking, the U.S. government may be able to keep its borders even more open to Arabs and other groups, knowing that if terrorist suspects slip through the cracks, authorities will have an easier time locating them.

Homeland IDs and Law Enforcement

A Homeland ID will make carrying out nefarious activities difficult not just for foreign visitors, but for U.S. citizens as well. Criminals will find that skirting the law and avoiding responsibility for their actions will become a tricky endeavor. For instance, consider the shortcomings in conducting a background check with the current system of identification. Because many background checks require a county-by-county search to uncover criminal activity, someone with a criminal record can easily provide a fake address to avoid being found. These loopholes have caused background checks for gun purchases to fail miserably. A story on National Public Radio (NPR) showed how easy it was for criminals and terrorists to buy guns in this country.[12] According to the report, some terrorists are buying guns in this country to ship overseas. A GAO report confirmed these findings and showed how simple it is for undercover agents to buy firearms using fake IDs.[13]

As a side note, the NPR story demonstrated the hypocrisy that is often found in the privacy debate. Here was a story based on a PBS show called "NOW with Bill Moyers" that frequently supports causes like civil liberties and the fight for privacy. The story was critical of John Ashcroft, who has required searches of voter registration lists, immigration lists, and driver's license records for potential terrorists, but not of the National Instant Criminal Background Check System (NICS) database. This database, created by the *Brady Handgun Violence Prevention Act* of 1994, requires that background checks be performed on people buying firearms.

In an amazing flip of positions, Ashcroft, with support from the National Rifle Association (NRA) argued for privacy and against searching the NICS database, whereas the civil libertarians from NOW were arguing against privacy and demanding that the gun list be opened. The government's hypocrisy is revealed when it asks for surveillance powers, except when they affect an important constituency, such as the gun lobby. Privacy advocates appear just as disingenuous when they demand privacy, except for a group they are opposed to, such as the NRA. Regardless of the politics involved, the NICS database should be opened and compared against the list of potential terrorists because guns greatly affect the security of our society.

A Homeland ID would raise the bar for a host of miscreants wanting to engage in criminal activity. It would make it much easier to track down people who commit fraud, individuals who skip out on trial, and parents who run off with their children. It would help prevent dangerous people, such as rapists and child abusers, from being hired into jobs were they could endanger others or from moving anonymously into unsuspecting communities where they could be a risk to children.

For instance, it was reported in 2003 that California lost track of thirty-three thousand convicted sex offenders who failed to report their addresses as required by Megan's Law. With a Homeland ID, the first time an offender failed to report, that person would be added to an offender watch list. The next time he or she used a Homeland ID during a transaction, authorities would be alerted.

Using biometrics to track down criminals is nothing new. For instance, consider the high-profile case of the D.C. snipers who in 2002 terrorized the citizens of several states, eventually killing ten people and leaving three wounded. It wasn't until the snipers bragged about a crime in Montgomery, Alabama, that investigators could match a fingerprint found at that crime scene with another stored in an immigration database under the name Lee Malvo.

It was ironic that with all their effort to avoid capture, the snipers were foiled by a simple biometric measure. It was equally ironic to see many of those in the national press in D.C., who typically argue for stronger privacy measures, praising the fact that there was a database containing Lee Malvo's fingerprint. Maybe living in the line of fire gives one a different perspective on the security/privacy debate.

Risk of Totalitarianism

Another argument against a Homeland ID is that it would lead to a totalitarian government that would monitor citizens and restrict movement. Although their causes may be numerous, more often than not totalitarian regimes arise out of social disruption and upheaval. Etzioni makes this point, suggesting that although totalitarian governments can use ID cards to repress their populations, they certainly aren't created by them. In fact, ID cards can help prevent the social chaos that would lead to the creation of one. Etzioni says,

> Moreover, libertarian concerns about totalitarianism confuse cause with consequence. Although ID cards can be utilized by totalitarian governments to restrict freedoms, these cards do not transform democratic societies into totalitarian ones. Totalitarian governments do not creep up on the tails of measures such as ID cards; they arise in response to breakdowns in social order, when basic human needs, such as public safety and work opportunities are grossly neglected. When a society does not take steps to prevent major social ills and strengthen law and order, an increasing number of citizens demand strong-armed authorities to restore law and order. By helping to sustain law and order, universal identifiers may thus play a role in curbing the type of breakdown in social order that can lead to totalitarianism.[14]

Although there have been isolated cases of abuse in various countries that use national IDs, their use hasn't led any of the democracies using them to the brink of dictatorship. It's also a non sequitur to say that because a regime like the Soviet Union used identification to monitor people and restrict their movements, a free and open democracy like the United States would do the same if it had a national ID.

It is true that occasionally people will be asked to show their IDs if police have reason to be suspicious. Police are allowed to do this today under the so-called Terry Laws, named for a 1968 Supreme Court decision in *Terry v. Ohio*. In fact, civil libertarians are up in arms over this in Wilmington, Delaware, where police are using Terry Laws to compile pictures of young males hanging out in seedy areas. Suggesting the "if you haven't done anything wrong, you don't need to be worried" principle, Mayor James Baker said, "Good little kiddies in the wrong place at the wrong time are not getting their picture taken."

However, in many cases the courts have protected the rights of individuals who refuse to turn over their IDs. In 2001, in the case of *James Carey v. Nevada Gaming Control Board*, the court found that agents for the gaming board had acted unconstitutionally when they arrested Carey, a suspected gambler, for refusing to turn over his ID. The court claimed that the arrest violated his Fourth Amendment protections. There is no reason to think that a Homeland ID would change the court's position on this matter.[15]

Discrimination and Harassment

Many people fear that minorities will face harassment and discrimination from authorities if a Homeland ID becomes the standard for identification. They point to cases where law enforcement has used racial profiling to determine whom to pull over for traffic stops. Although these incidents are intolerable, it is reassuring that when they have been uncovered, they have been dealt with swiftly and justly. For instance, in several states the Justice Department reached agreements with police departments to end the practice, and on the federal level, the Bush administration has issued a ban on using race or ethnicity in routine criminal investigations. Unlike totalitarian dictatorships, open democracies like the United States tend to have good records for sniffing out such abuses and eventually correcting them.

A related concern is that Muslim Americans might be detained at airports and other security points if officials happen to confuse them with terrorists. An article in *Wired News* entitled "Due Process Vanishes in Thin Air" describes the case of Asif Iqbal, who, because he shares a name with a suspected terrorist detainee at the Guantanamo Naval Base in Cuba, has to get FBI clearance to fly each week.[16] Issues like these demonstrate the complete and utter failure of the current identification system to distinguish effectively between people. If a Homeland ID were combined with a biometric measure, the terrorist watch list would work as designed, and there would be no confusion between Asif the patriotic American and Asif the al Qaeda terrorist.

Paradoxically, a Homeland ID might lead to reduced profiling. Alan Dershowitz, who supports the idea of a national ID card, has made this point: "Four Arab-looking guys reading the Koran are much less suspicious if they have the cards and can just slash them through card readers."[17]

This points to the reason technology can enable Americans to have both security and freedom. Instead of relying solely on the biased perceptions of officials who often make judgments based on factors like skin color, authorities can use technology to make improved estimations of who is a threat. For the 99.99 percent of Americans who are not on a watch list, secure identification will greatly expedite security procedures. We see evidence of this with the INS Passenger Accelerated Service System (INPASS). Used at a handful of U.S. international airports, the voluntary system scans the hands of travelers who have gone through an enrollment process. Once the machine verifies their handprint, an automated gate opens and the traveler is allowed to skip long queues of people waiting to be interviewed by an immigration inspector. Instead of casting a wide net around thousands of innocent people, identification systems like INPASS and U.S. VISIT enable authorities to target precisely the minute number of people who are truly dangerous.

This leads to the surprising conclusion: The absence of a secure system of identification in the United States can be a greater threat to freedom at times. The case of a serial killer in Louisiana demonstrates this point. For over a decade, Derrick Todd Lee successfully covered up the trail of women that he had raped and murdered, giving authorities few leads other than leaving his DNA as a calling card.

In a more open society, where a person's DNA, rather than being shrouded in privacy, was available as an identifier, the killer's unique genetic sequence would have given him away long ago. James Watson, the codiscoverer of the DNA double helix, suggested during an interview marking the fiftieth anniversary of the discovery that governments should take DNA samples of everyone at birth.[18] The DNA would be stored in an international database to help fight crime and terrorism and to absolve individuals who are falsely accused. Furthermore, because DNA is the same for every cell in the human body, it would be more difficult for terrorists or criminals to circumvent.

The use of DNA databases is well established in the United States. The military stores DNA samples from all enlisted servicemen and women to help with identification if they are killed in duty. Howard University is using a gene repository for African Americans to help identify genes involved in diseases that primarily affect blacks. All fifty states collect DNA samples from convicted felons and sexual offenders for forensic purposes. Virginia, which set up the first such database in 1989, claims that it has helped law enforcement officials solve more than 100 homicides, 200 rapes, and 450 burglaries.[19]

Look at the alternative to secure identification. In Louisiana, authorities used the relic technique of racial profiling to search for Lee. Although racial profiling is an appalling way to investigate a case, investigators weren't even able to get the race correct, believing the killer to be a white male. Authorities held the key to Lee's identity, his DNA, but because they had no good way of using it, Lee was able to snuff out the lives of more unfortunate women.

Furthermore, due to anonymous tips that a white truck was seen near the crime scenes, the hunt for Lee resulted in a dragnet targeting the owners of the twenty-seven thousand white trucks registered in Louisiana at the time. One could suggest that using such a feeble link to the crime to pull over drivers was an infringement of citizens' liberties. It was at least a significant harassment for large numbers of citizens who had to endure regular stops by police when their white trucks were spotted on the road. Many people are so busy fretting about a national ID that they fail to see that time and time again it's the lack of identification that leads to the targeting of innocent individuals and the restriction of their liberties.

In addition, as Lee roamed free, women throughout the state were lining up to buy guns and pepper spray, while others were afraid to leave their homes. Arguments that claim an open society must accept a certain amount of crime fail to recognize that individual liberties are often sacrificed when an individual's safety and security are forfeited.

In a technologically advanced country dedicated to protecting civil liberties like the United States, we can reject notions like the Devil's Dichotomy. Technologies of openness like the Homeland ID allow us to have our cake and eat it too. We can be both free and secure.

Slippery Slopes of Abuse

Another concern with a Homeland ID is that it would lead America down a slippery slope of abuse. Many point to how the SSN was originally designed for the government's retirement program, but eventually expanded to a variety of uses in the public and private sector. Although a statement of fact, this argument fails to mention that multiple applications of the SSN resulted from the significant need in a modern information society to identify people. Although the SSN hasn't worked very well in this capacity, there have not been many decent alternatives. Could you imagine life where people had to keep track of ten, twenty, or thirty numbers instead of one SSN? People have enough trouble remembering a single password in the corporate world. A Homeland ID would eliminate this problem. The SSN could still be used as a key to store records in a database, but if you wanted to access your record, a biometric measure like a fingerprint would serve as the password.

The other component of the slippery slope criticism is that a Homeland ID would lead to a centralized database of information on individuals that could be misused by authorities. Although the *Federal Privacy Act* of 1974 and the *Computer Matching Act* of 1988 place limits on the sharing of information between government agencies, in many cases, such as in tracking down deadbeat parents, agencies have gotten around these restrictions.

Although this chapter argues that a Homeland ID can be implemented without resorting to a centralized database, there will be continued pressure for the government to link to other databases, especially when it can do so more effectively with a national identifier for individuals. For instance, with burgeoning budget deficits on their hands, state governments are linking departments and their databases to track down tax evaders. For example, if a taxpayer reports a small income to the state, but DMV records indicate that he or she owns an expensive car, the taxpayer's record might be flagged for an investigation.

Many of those who would like to keep the government from participating in the information age of data sharing see justification in the argument that government inefficiency serves as an important role in our constitutional system of checks and balances. In *Justice Dept. v. Reporters Committee* (1989) the U.S. Supreme Court argued, "There is a vast difference between public records that might be found after a diligent search of courthouse files, county archives, and local police stations throughout the country and a computerized summary located in a single clearinghouse of information."[20]

Although this argument may have made sense to the Founding Fathers when communications could take weeks to arrive and government inaction had minimal impact, the consequences today of the failure to react immediately to a threat could mean the difference in stopping the spread of a smallpox attack or preventing the detonation of a dirty bomb. Furthermore, although the court makes a valid argument about the difference between card files in a courthouse and digital records in a single clearinghouse, as more e-government initiatives are introduced, we will soon arrive at a time when all records are digital and such a distinction is no longer tenable.

Instead of backwards-looking arguments supporting government inefficiency, the focus should be on policies that shine the light of transparency on those who have access to the information. As Chapter 7 shows, stove-piped, or isolated sources of, information contributed to preventing U.S. intelligence from connecting the dots before 9/11. The question should be not how we can keep information out of the government's hands, but how we can guarantee that the government uses it properly.

Surprisingly, given all the hoopla over government databases, one doesn't read of many cases of rampant abuse. There certainly are some horror stories, such as when five hundred IRS employees were caught snooping through the files of their neighbors in 1995. However, as is the case in most democratic societies like the United States, clear examples of misuse that outrage the public are usually corrected over time. For instance, in this case, Congress passed the 1997 *Taxpayer Browsing Protection Act* to prevent IRS workers from snooping into Americans' tax records.

In a day and age when anyone can do a search on Google to uncover a flood of information about a person and when private companies routinely match information from hundreds of databases to build detailed profiles of consumers with more data than the government could ever hope to access, some of these fears seem a little outdated. As Robert Kuttner argued back in 1993 in the *Washington Post*:

> The idea that any of us is sheltered from countless national data bases or ID cards has long since been overtaken by technology. Americans are already vulnerable to massive invasions of their privacy, courtesy of computerized data bases and ID cards. The paradox of our national phobia against ID cards is that we already have most of the liabilities while denying ourselves potential benefits of computerized record keeping.[21]

Despite fears of national identification, it's likely that the United States will eventually be forced to move to some type of improved system. If identity theft continues to rise at its current rate and affect all strata of society, there's a good chance that the public will demand that the U.S. government make changes. If there is another attack by terrorists who use counterfeit IDs to thwart authorities, the pressure on the government to create a Homeland ID might be irrepressible.

As more attention is focused on the issue, it is more likely that valid concerns will begin to be separated from overblown rhetoric like wheat from chaff. As the facts are revealed, people may begin to question whether they really have to make a choice between freedom and security. Economist James Glassman puts the question this way:

> Dangerous? Yes, there are dangers to a mandatory national ID card, but there may be greater dangers without one. The fact is, to live in a society as vulnerable as ours, we may have to give up something—but I disagree that what's lost is freedom. Instead, it's privacy, and maybe not even that. . . . The truth is that an ID card may force you to give up some of your privacy—though probably no more than driver's licenses, Social Security cards, credit cards and even electronic toll-readers like FasTrak force you to give up now. But even if privacy is lost, the question is whether such an exchange is worth the benefits? More and more, I believe it is.[22]

Malcolm Anderson, a professor of politics at Edinburgh University who has studied the issue, discounts the exaggerated concerns over national identification. He says he is "dubious about suggestions that ID cards would be an infringement of civil liberties. The only argument that I can take seriously is that you lose the right to disappear."[23]

In a more open society, where identification is no longer hidden in the murky waters of anonymity, the loss of the right to disappear will resound the loudest among the ill-intentioned, who will find there are very few shadows left in which to lurk.

CHAPTER 6

SMILE, YOU'RE ON CANDID CAMERA

The Case for Surveillance

Of all the trends from the technological revolution of the last several decades, the growth in surveillance has been one of the steadiest and most consistent. Ever since the early 1960s, when federal law mandated the use of video cameras in banks, surveillance in all its many forms, from satellites to sensors to facial-recognition cameras, has become increasingly woven into the fabric of our everyday lives.

The quickest way to get a sense of the extent of what has been called the surveillance society is to consider the proliferation of cameras in America. According to one estimate, there are two million of these devices in operation, and the number is growing fast.[1] Revenue for the video-surveillance industry, which was $282 million in 1990, is predicted by the Security Industry Association to grow to $1.63 billion by 2005.[2] Prior to 9/11, a count by the ACLU of cameras in Manhattan revealed 2,397 of the devices.[3] After 9/11, one group estimated a 40 percent increase in cameras in that same area.[4]

Cameras are not just increasing in numbers, but in their power and capability. The latest gizmos provide sharp color images that use tilting, panning, zooming, infrared, and motion-detection technology and store images in digital format for easy computer access. New systems can create a clear picture at thousands of feet, compared to a few dozen feet with older cameras, and transmit the video over wireless networks to portable devices such as Palm Pilots or laptops. Paired with video-processing computer chips, these various devices are becoming intelligent, gaining the ability to determine when a human enters the camera's field of vision, recognize the subject's face, and then follow the subject's movements.

The future of surveillance cameras as described in science fiction books is now becoming a reality. For example, there are cameras that can see through walls and clothing by creating an image from the ultra high-frequency energy naturally emitted by objects. Other smart cameras can, when arrayed together in a network, communicate with one another to track an object as it moves from one field of view to another. Some companies are working on systems that can create a 3-D model of an area and overlay it with video feeds from many cameras, making monitoring as easy as maneuvering a joystick.

Although the sheer numbers of sophisticated cameras and other recording devices is striking, the integration of these systems into larger networks promises to drive the real revolution in surveillance. For example, in Washington, D.C., police are using a multimillion-dollar Joint Operations Command Center located on the fifth floor of their headquarters as the foundation for a nascent surveillance infrastructure in the nation's capital. The center has a wall of video screens that will allow law enforcement and even the FBI and Secret Service (both of which have reserved workspace) to view feeds from local schools, intersections, the metro system, and other sites deemed important. This video hub may eventually serve as a model for other communities, foreshadowing a time when cities and towns across the country are linked together in a network of surveillance.

But before we get into a fretful discussion about life with an omnipresent eye in the sky, it might be helpful to put the debate into perspective by examining a side of surveillance not often mentioned—how it helps contribute to the public's health, safety, and welfare.

Everyday Surveillance to Promote Safety

Although much angst has been expressed over the rapid spread of surveillance and its potential for abuse, in a number of other less publicized instances, cameras are innocuously designed to promote the public's health and safety. For example, take the use of video devices in schools. Since the Columbine shootings, when it was discovered that police were hampered by not knowing the location of the shooters, many school administrators and law enforcement officials have installed cameras in hallways, stairwells, parking lots, and other common places where security is known to be a problem. These video systems allow police and fire officials to log in remotely at any hour of the day and get a number of different views inside and outside the school.

Watching over children during their free time, such as during a day at the pool, is another difficult task made easier by surveillance. One company uses cameras above a pool and below the water to record activity, which is analyzed by a central computer. The system's algorithms can identify swimmers who become still for more than a few seconds, sound alarms, and use waterproof pagers to notify lifeguards.[5]

Cameras are being employed in a variety of health-care environments, nursing homes, and daycare centers to assure that the proper care is given to clients. For example, many companies are working on home surveillance systems that monitor the elderly.[6] Through the use of motion detectors on walls and in cabinets, sensors on pill boxes and plates, and surveillance cameras throughout the house, a computer containing special algorithms can determine whether a person is eating and taking pills on a regular basis or whether the person has fallen down or been motionless for a certain period of time, indicating a possible emergency.

In the journal *Pediatrics*, a study showed that hidden surveillance devices were effective in uncovering cases of Munchausen's Syndrome by Proxy (MSBP). This syndrome is a type of abuse where parents intentionally poison their children with toxic substances to keep them sick. Because the parents are skilled at deception, doctors often have a hard time discovering the true cause of the illness. In the study, the researchers recommended that all children's hospitals have cameras in place to monitor parents in suspected cases of MSBP.[7]

Parents worried about their children's safety can turn to Global Positioning System (GPS) bracelets to track their whereabouts. The child can dial 911 from his or her wrist, and if someone tries to remove the bracelet forcefully from the child, it automatically calls the police. In the near future an implanted chip may replace the bulky bracelet. Applied Digital Solutions offers the Digital Angel, a rice-sized chip that, when placed under the skin, wirelessly transmits an ID number or other information to a scanner. The controversial product, which is already approved by the FDA, costs $200.

Victims of traffic accidents can benefit from these types of technologies. OnStar, GM's onboard communications system, offers a GPS service to millions of its customers. If you are in an accident and the airbags deploy, OnStar can alert nearby emergency services to your location, even if you are unconscious. Event data recorders under the hoods of many vehicles take safety a step further by determining the severity of a crash and relaying this information to a central office, which can call paramedics.

In the event that your car is stolen or hijacked, systems such as OnStar can thwart the thief's efforts by disabling the engine or allowing authorities to pinpoint the automobile's location. Car dealers are leveraging this technology when they extend credit to consumers who are bad credit risks. If a payment is late, the dealer can remotely shut off the car. Many drunk drivers are intimately aware of such systems as some states have forced breath locks onto cars, forcing these high-risk drivers to breathe into a breathalyzer before the engine can start.

A modification of the event data recorder is being marketed to parents who can use it to download information about their teenager's driving habits. The system can even be set up to emit loud beeping sounds if the child begins driving too aggressively. Although the question of whether a spouse, employer, or law enforcement official will be able to use such tracking devices has yet to be determined by the courts, it's not likely that the legal system will have much sympathy for the philandering husband caught with his pants down.

Enforcing Traffic Laws

Public safety can be enhanced further when surveillance allows authorities to better enforce the law, particularly when it comes to traffic violations. In many areas, limited coverage by police means that drivers are more likely to speed or run red lights, thus increasing the chance of traffic accidents. However, as cameras begin to extend the reach of law enforcement, reckless drivers are learning that violating traffic laws is an expensive endeavor. I found this out recently after a picture of my car going through a red light was waiting in the mail for me with a fine courtesy of the city of Fairfax, Virginia.

In the United Kingdom, where cameras have become as common as trees along the side of the road, police are using surveillance to catch people who litter from their cars. The technology, which recognizes license plates, allows authorities to identify stolen cars or ticket people with expired registrations. Taken a step further, the country is experimenting with automobile surveillance to control traffic into cities, for instance by charging commuters higher rates for traveling during peak hours.

Some cameras can even recognize an accident and alert authorities. Germany is installing smart systems in tunnels to identify smoke and fire from accidents.[8] The cameras have computer chips that can tell the difference between truck exhaust and smoke from car fires.

Is there any evidence that roadside cameras can help reduce the number of traffic accidents? According to CamerasCutCrashes.com, the first year of a pilot study found that crashes fell by 65 percent at new camera sites, and deaths and serious injuries dropped 90 percent. A Charlotte, North Carolina–based study conducted over a three-year period concluded that the number of front-to-side crashes at intersections equipped with red-light cameras decreased by 37 percent, even though rear-end collisions increased by 4 percent.[9] This is good news because front-to-side crashes are much more dangerous.

Fighting Crime

One of the first recorded cases of police using cameras for surveillance was in Hoboken, New Jersey, in 1966, although the system was dismantled after little evidence of its success in fighting crime.[10] Since that initial experiment, advances in technology have made surveillance much more appealing to law enforcement.

It's well known that the United Kingdom, with a network of several million closed-circuit televisions (CCTV), is the undisputed leader in using surveillance to fight crime. Recent reports suggest that the United States may be closing in on their lead. A 2001 survey by the International Association of Chiefs of Police reported that 80 percent of nineteen thousand U.S. police departments are using closed-circuit television, and over the years there have been some notable results.[11] A bank ATM camera filmed the Ryder truck outside Oklahoma City's federal office building just before the blast in 1994 that killed 167 people. That clue helped police track down Timothy McVeigh.

Police will even turn to a private surveillance system if they think it can support their law enforcement efforts. During the hunt for the D.C. sniper, officials worked with local businesses to identify video footage that could provide clues to the shootings. Unfortunately, many times the camera was pointing toward the target of the shooting and not toward the snipers.[12]

Law enforcement officials certainly can use all the help they can get when going after criminals. As we saw with the statistics provided by Brin, out of every one hundred felonies committed, only a handful are ever successfully prosecuted. An officer interviewed by Richard Ericson and Kevin Haggerty said,

> A: Do you have any idea what our clearance rate for break and enters was last year? Three per cent! We have almost no opportunity to catch people breaking into homes unless we just happen to stumble across them.
> Q: Or unless they run into the side of your car?
> A: They would have to run right into an open door and into the back seat for us to catch them.[13]

Although not a cure-all for ending crime, cameras and other surveillance devices can play a part in an overall plan to combat it. Consider the number of different crimes that happen in public. People get mugged. Cars get stolen. Houses get broken into. Kids get kidnapped. When there are limited numbers of police on the street, criminals are smart enough to know that they have a reasonable chance of getting away with their illicit deeds. A camera changes the equation and makes criminals think twice about whether video devices might help identify them at a later date.

Is there any evidence that cameras are effective in helping to fight crime? So far, the limited research on the issue has been mixed. Some studies show decreases in crime. In one independent evaluation of a CCTV program in Hull, England, car crime dropped 80 percent; shoplifting, 69 percent; robbery, 68 percent; burglary, 49 percent; and violent crime, 30 percent.[14] According to the Downtown Partnership of Baltimore, Maryland, a study of sixty-four cameras reported that crime was reduced by 15 percent.[15] In New York, former mayor Rudolph Giuliani claimed that surveillance reduced crime in public housing by 20 to 40 percent.[16] In other cases, the results haven't been so spectacular.[17]

Some argue that because determined criminals adapt to the state of law enforcement, efforts to impede them lead to an arms race. For instance, some studies have shown that once cameras are installed, crime can rebound after an initial drop or move to other areas. According to one report, criminals in England try to avoid surveillance by stealing cars on the move or embracing victims during a mugging to make it appear to be a romantic moment.[18] Of course, one might ask, if cameras don't deter crime, why do so many private companies use them for security. The private sector generally doesn't continue to invest in something unless it offers some measure of benefit.

Surveillance in the War against Terrorism

As is apparent from the examples listed above, the number of applications of cameras and other monitoring devices is growing rapidly. Since 9/11, surveillance may be asked to play another role, securing the homeland and protecting critical infrastructure. Consider the gargantuan task of guarding an endless number of vulnerable sites around the country from attack:

> The Golden Gate is, after all, just one of 590,984 bridges around the nation. There is one Hoover Dam, but 54,065 public and private water systems. Eighty-five deep-draft ports. One hundred and three nuclear power plants. Untold miles of highways, railroads, underground tunnels and oil pipelines, innumerable electricity grids and telecommunications hubs, each vulnerable to attacks with the potential to disrupt commerce if not endanger lives.[19]

Some cities have recognized the vulnerability of their infrastructures and are turning to surveillance cameras for assistance. The California Department of Transportation installed video cameras in the Bay Area to guard transportation infrastructure like the Golden Gate and Bay bridges. The video is sent wirelessly to officials and engineers at the department, leading to huge savings by eliminating expensive wiring and reducing dependence on human resources.

Reliance on humans is further lessened when video surveillance is combined with powerful computer systems. For example, the Unattended Baggage Detection System from NiceSystems can analyze a snapshot of an area in an airport and detect whether an object such as a bag has been left, something airport officials do manually today.[20] ExitSentry is a product that can spot people trying to enter through the exit lane of an airport. Cernium markets the product as a way for airports "to replace the costly FAA-mandated live guard within exit lanes in airport concourses."[21] More than a dozen airports are currently using the technology. Another company, Equator, builds computer chips into cameras that allow them to detect whether two people walk through a door after only one person swipes a badge.

Intelligent surveillance technologies smart enough to identify the shape or face of a human or detect abnormal behavior such as someone dropping a package in a secure area can be deployed to watch over bridges and power plants, electric grids and pipelines, national monuments, skyscrapers, and public facilities. For instance, radar technology being developed by the Defense Advanced Research Projects Agency (DARPA) that can uniquely identify a person by the way he or she walks could be used to alert security guards to an approaching figure in the dark of night who wasn't a regular employee. The system can even detect if a large package, such as a bomb, is being carried on the person's back.

Another Herculean challenge is watching over national borders and ports. The U.S. border with Canada spans 3,897 miles and that with Mexico spans 1,933 miles. Combined with coasts and islands, the border totals nearly 9,600 miles.[22] The costs to secure this area through traditional means such as helicopter or foot patrols are overwhelming and, consequently, are limited to high-traffic areas. Through the use of pilotless Unmanned Aerial Vehicles (UAVs), like the Predator, which saw success in military campaigns in Iraq and Afghanistan, the U.S. government can greatly increase its ability to look for people slipping into the country.

The task of protecting U.S. ports is just as overwhelming. Currently, the government screens less than 2 percent of the twenty-one thousand containers that arrive at ports each day.[23] Forty feet long, these containers contain enough space to hold the contents of a house.

In one disquieting scenario, a terrorist uses a shipping container to smuggle nuclear material for a dirty bomb into the United States. To forestall this possibility, U.S. Customs agents are deploying an array of new technologies that can effectively see through boxes and crates for nuclear contraband. One product employs a truck-mounted cesium or cobalt detection system. The system uses hundreds of sensors that send radiation through the container and then convert it into a picture that officials can use to identify high-density material of the nuclear kind.

Other emerging technologies enable authorities to detect deadly agents that might be released in the nation's cities. For example, the Bush administration is rolling out a nationwide pathogen-detection system that uses the Environmental Protection Agency's (EPA) three thousand air-quality monitoring stations to sample for biological or chemical agents. In Washington, D.C., and surrounding communities, the government has been installing thirty-foot-tall towers on top of government buildings and other important sites; these towers can detect radiation from a nuclear attack and predict which way the fallout will blow.

Obviously, the United States will never have enough manpower to watch over all of its most critical assets, especially during times of heightened terrorist threat. According to some estimates, up to 80 percent of the nation's critical infrastructure is in the hands of the private sector. American companies are unlikely to be able to bear the burden of security, and with budget shortfalls, support from local, state, and federal governments is likely to be limited.

There are a number of ways to compensate for these deficits, especially during periods of increased threat, including forcing law enforcement officials to work overtime, calling up the National Guard, shutting down public buildings and monuments as was done immediately after 9/11, or distributing security resources based on risk as the National Targeting Center currently does at U.S. ports. Although these efforts may provide a quick fix to specific vulnerabilities, none offers a long-term, comprehensive, and satisfactory answer to the question of how best to secure the homeland.

In an open society like the United States where a wide range of potential targets exists, it is unrealistic to think we can guard against all of the different possibilities of attack. However, the increasing sophistication and falling cost of surveillance technology gives America the opportunity to extend dramatically a widening blanket of security over important national assets. A strategy to protect critical infrastructure that consists of sensors, detectors, cameras, and other devices will enable the country to use its human resources better while raising the bar for those who would carry out sinister plots on U.S. soil.

Facial Recognition

Perhaps the most significant advance in surveillance technology, and one that may prove the most useful in the war against terrorism, is facial recognition. As was mentioned above, this biometric technology compares the digital pattern of a human face against others in a database in order to find a match. For instance, the University of Missouri, Rolla, has a research facility that uses a 200-kW nuclear reactor with low-enriched uranium to train nuclear engineers.[24] Although reactors at research facilities are smaller than those at nuclear plants, there is still the threat that the nuclear material could be stolen for use in a dirty bomb. As a result, the university has implemented a facial-recognition security system as a way to limit access to the facility.

One of the first public uses of such a system was during Super Bowl XXXV in Tampa Bay, Florida, where police used an application called FaceIt to scan the crowd for the faces of criminals. Shortly thereafter, the city took face recognition one step further and became the first to use the technology on public streets. Civil libertarians were outraged by Tampa Bay's actions. The ACLU of Florida said in a letter to city officials, "While everyone has a reduced expectation of privacy while in public, including sitting in the stands with one's family at a Sunday afternoon football game, we do not believe that the public understands or accepts that they will be subjected to a computerized police lineup."[25] It's also likely that the public understands that there is little privacy for individuals' faces in public, whether they pass a policeman on the street, a camera overlooking a building, or a hotdog vendor in a stadium.

Although Tampa Bay eventually ended its experiment with facial recognition, other cities have continued to explore its uses. For instance, some school systems are experimenting with systems that compare the facial patterns of visitors against databases of known child molesters. If a match is found, a human verifies the results before any action is taken.

Perhaps the best use for a technology still in its infancy is found in more controlled environments with a specific security need. In the fight against terrorism, facial recognition can be deployed to provide additional assistance in watching over important gates, such as subway systems, building entries, and airport terminals. As efforts to photograph foreign visitors to the United States expand, watch lists will eventually have a larger pool of pictures to draw from as new terrorist suspects are identified and targeted by authorities. Used in this capacity, facial recognition can compensate for occasions when terrorists obtain counterfeit identification in an effort to avoid having their names matched to a watch list.

Surveillance of Communications

As we saw earlier, in a decentralized organization like al Qaeda, there is great dependence upon communication in order to stay connected. Even though the group works hard to keep their messages secret by using encryption, code words, and other techniques, the sophisticated surveillance capabilities of the United States and its allies have been critical thus far in identifying and capturing members of sleeper cells around the world. The buzzword "chatter" has been popularized since 9/11 to refer to the gathering of raw intelligence data on terrorists groups, including eavesdropping on communications. Intercepts of satellite phone transmissions were recognized as the main reason the

United States was able to keep tabs on Osama bin Laden for such a long time, at least until word leaked out about the operation.

One system that the United States uses for monitoring chatter is Echelon, a secret electronic eavesdropping system managed by the intelligence organizations of the United States, United Kingdom, Canada, Australia, and New Zealand. Using land-based antennas, satellites, taps into fiber-optic cables and other means, the system gathers large amounts of e-mail, telephone, fax, and other types of communication data that are then sifted using keywords deemed suspicious. Very little is known about Echelon, and the countries involved have consistently denied its existence.

Another communications surveillance program is known as Carnivore. Carnivore, which was renamed DCS 1000, is a system the FBI uses to search Internet and e-mail traffic. The FBI installs Windows-based computers by court order at Internet service providers, which it then uses to sift through e-mails. This type of filtering can save time for FBI agents by helping them focus on the most relevant e-mails in what can be an overwhelming flood of messages.

Although still a secretive program, more information is available about Carnivore than Echelon. A report authored by Henry H. Perritt Jr., dean of the Chicago-Kent College of Law, that studied Carnivore found its data-collection practices reasonable and in line with what the FBI had claimed, although critics have attacked this report as biased.[26]

Legal Limits on Government Surveillance

The U.S. government is often the locus of the fears of those who worry about the proliferation of surveillance. Although many of the technologies are created and managed by the private sector, concern with the government is justified because it has a monopoly on the lawful use of force. Moreover, the public is much too familiar with episodes in history where the government misused its powers in order to spy on Americans.

Although these concerns are reasonable, one must not forget that partly as a result of past abuses and partly as a result of checks and balances in the Constitution, there are a number of legal limits placed on the government's power in this area. One of the first places to find these limits is in the Fourth Amendment.

The Fourth Amendment serves as the Constitution's primary mechanism for protecting the public's privacy from invasions by the government. The amendment states,

> The right of the people to be secure in their persons, houses, papers, and effects, against unreasonable searches and seizures, shall not be violated, and no Warrants shall issue, but upon probable cause, supported by Oath or affirmation, and particularly describing the place to be searched and the persons or things to be seized.

The Fourth Amendment restricts the government in two particular actions: searches and seizures. Although a "seizure" refers to taking control of property or arresting a person, it is a "search," which the courts have defined as an invasion of a person's reasonable expectation of privacy, that has ramifications for surveillance.

As Jeffrey Smith and Elizabeth Howe point out, although the Fourth Amendment places limits on the government's ability to carry out a search, it doesn't prevent it. Furthermore, it doesn't address surveillance in the public setting, which, according to the U.S. Supreme Court, does not even count as a search.[27] If the government wants to engage in surveillance for a law enforcement purpose, it may do so as long as the surveillance doesn't violate a person's reasonable expectation of privacy; if it does, a court may grant a warrant after being satisfied that there is probable cause that a crime has been or is being committed. The exceptions to the rules are for searches that occur as part of a lawful arrest; are carried out in particular places, such as an airport or sobriety checkpoint; or involve emergency situations, such as fire or safety.

In 1967 in the case of *Katz v. United States*, the Court determined that the surveillance of communications qualified as a search under the Fourth Amendment. A year later, Congress extended the Fourth Amendment with the *Omnibus Crime Control and Safe Streets Act* of 1968, known as the *Wiretap Act*, Title III of which defined the standards for monitoring communications. In 1986 Congress took these protections a step further when it passed the *Electronic Communications Privacy Act* (ECPA) to include computer communications. Although this statutory framework does not offer a high level of protection for "envelope" information, such as the addressee on an envelope or of an e-mail, content information, such as the text of a letter or e-mail, is strictly safeguarded by a requirement of probable cause. Ironically, as Orin Kerr points out, the *USA PATRIOT Act* was criticized for changing the standards for Internet envelope information, when in fact, it strengthened protections by affirming that this kind of information requires a court order, something that was unclear in the past.[28]

Compared to the electronic monitoring of the contents of communications, protections against video surveillance are usually much lower; in many cases they are nonexistent. In circumstances where cameras or facial-recognition systems record actions in public and are not directed at a specific person, a warrant is not typically required, and the criterion of reasonable expectation of privacy is used to judge what is permissible. Since *Katz v. United States*, courts have consistently found that there is little expectation of privacy in public and that most spaces available to the public eye are not protected by the Constitution; for example, one has no expectation of privacy concerning one's garbage (*California v. Greenwood*, 1988)[29] or in one's travels on the highway (*United States v. Karo*, 1984).[30] This contrasts with privacy in the home, which was affirmed when the Supreme Court ruled in the case of *Kyllo v. United States* (2001) that police using a thermal imaging device on a suspected marijuana grower's home violated the Fourth Amendment's restrictions against unreasonable searches and seizures.

Jeffrey Rosen suggests that cases like *Kyllo* indicate that the courts are using the boundaries of the home to demarcate private space protected by the Constitution from public space that is not.

> If you were forced to generalize, it would be the unhappy truth that there's not a constitutional or statutory protection against surveillance in public places. Courts have held in a lot of contexts that you have no expectation of privacy in places where you voluntarily expose yourself to the world. And they've drawn a distinction between the home, which does have special protection against cutting-edge surveillance technologies, and public spaces, which tend not to.[31]

However, understanding the shifting boundaries of the limits to surveillance isn't as simple as drawing a line around the home. Consider the 1989 ruling in *Florida v. Riley*.[32] In this case, authorities received an anonymous tip that marijuana was being grown on private property. When the investigating officer couldn't view the contents of a greenhouse, the officer used a helicopter to fly over the property to get a better look. The Supreme Court in a 5–4 ruling concluded that the Fourth Amendment did not protect the owner's property because anyone flying over the house at a reasonable altitude would have been able to see the same thing. So even though the greenhouse was a part of the home, the courts found that because it was exposed to the public, the Constitution didn't provide a shield against surveillance.

Many advocates in a seemingly losing argument over constitutional protections for privacy in public spaces have redirected their fight against being observed to a fight against being identified. The reasoning appears to be that if one can't prevent cameras being placed in public, perhaps limits can be placed on smart cameras that can recognize people. Says Marc Rotenberg of EPIC,

> Do you have an expectation of privacy in a public place? A lot of people say, "Come on, don't be ridiculous, people can see you." Well, they may be able to see you, but they won't be able to recognize you. Observation and identification are two very different outcomes. There are clearly Fourth Amendment rules that prohibit people from routinely stopping people on the street and asking them for their identification papers. So, technology that makes possible identification is very different from one that merely allows observation.[33]

This position does not have much grounding in the Constitution and is unlikely to find favor in the courts, especially when the technology is no more intrusive than a person eyeballing a neighbor on the street is. Without wearing a bag over one's head, to use Alan Dershowitz's analogy, it is simply impossible to prevent identification from occurring when one is out in public. About the best hope for those worried about smart surveillance is to create a groundswell of support among the public to drive Congress to enact legislation to ban or regulate the technology. John Woodward, a former CIA operations officer and analyst with the Rand Corporation, echoes these sentiments:

> It is a feeling. It is not a legal right. When you're in public, you might think you are anonymous, but your friends recognize you if you're with another woman and your wife recognizes you, etc. There is a feeling. But a feeling doesn't equate to a legal right in my opinion. Maybe the argument is, hey, let's have public privacy. But again, that's why I say at the outset that's for politicians to decide. I don't think the courts are going to find it.[34]

Criticisms of Surveillance: Potential for Abuse

Criticism of surveillance and facial-recognition systems has been expressed on a number of grounds. One claim is that cameras are easily abused; for example, they enable lewd spying on others. One notorious example is of a plumbing supply company in St. Louis, Missouri, called Miracle Supply, where the CEO placed a surveillance camera in the women's bathroom. Another case involved male university athletes being filmed secretly in locker rooms by men posing as coaches or referees.[35]

Another concern is the use of surveillance to target blacks and other minorities unfairly. Critics point to a study by Hull University that showed minorities were more likely to be the focus of a camera. The study also found cases of camera operators leering at women.[36]

In researching this book, I found it to be true that there are plenty of outrageous stories of abuse that highlight the qualms expressed by opponents of surveillance. However, in light of the fact that there are millions of cameras in place around the United States, evidence showing systematic patterns of misuse is surprisingly sparse. Perhaps this is because more research is needed or because cases of wrongdoing do not easily make it into the public sphere. Nevertheless, until more evidence shows a frequency of misuse, one has to assume that a lot of anxiety is being generated unnecessarily from a handful of high-profile anecdotes.

In many of the cases where the law has been violated, such as with the Miracle Supply example, the perpetrators have been taken to court and prosecuted. Elsewhere, proactive steps have been taken to limit the likelihood of cameras being misused. For instance, in Virginia Beach, Virginia, police officials worked with local minority groups such as the NAACP to get their input on designing and overseeing a surveillance system.[37]

Lewis Maltby, director of the ACLU's task force on workplace rights, admitted that politicians are hesitant to restrict surveillance and acknowledged that in most circumstances it's not being used for malicious intent. He says, "Most secret taping isn't done by perverts. It's done by good guys—employers, landlords, police. When legislators discover their powerful constituents are taping, they back off."[38]

In a majority of situations, cameras are being used for very specific and practical purposes, such as guarding a parking lot or keeping an eye on kids at a daycare. Even a privacy advocate like Simson Garfinkel tells in his book *Database Nation* of setting up a camera to spy on his real estate agent.[39] It served the reasonable purpose of watching over his home and keeping his real estate agent accountable.

Making People Uncomfortable

Many argue that surveillance cameras make people feel uncomfortable and lead them to act unnaturally. This line of thought suggests that in front of the lens individuals will be more self-conscious and concerned about behaving within the norms of society. Jeffrey Rosen points to surveys showing that employees who are electronically monitored experience higher levels of depression, anxiety, and tension.[40]

Common sense says that it is likely that any employee constantly monitored by a manager will be a little stressed, camera or no camera. After all, this person affects the employee's ability to bring home a paycheck and put food on the table. This may be especially true for those employees accustomed to goofing off on the Internet or passing time with coworkers at the water cooler when the boss is away. The days of a free ride at the company's expense may be dwindling.

When it comes to cameras in public, the situation is a little different. How likely is it that a person behind the camera is going to recognize someone through the lens? It may even be the case that there isn't anyone manning the controls. I'd expect that of the millions of cameras in the United States, many, if not most, do not have a human being hidden behind the monitor.

Even in the case of facial-recognition systems, where the limitation of not having enough humans to watch every camera could eventually be overcome, most Americans are sensible enough to know that with hundreds, if not thousands, of people passing by cameras, they are just as unlikely to be bothered by an impersonal computer as they would be by a complete stranger on the street. John Woodward makes this point: "Once you leave your house and enter public spaces, just about everyone you can see can see you right back. . . . The bottom line is that most of us are very boring. We flatter ourselves to think that someone is building a multibillion-dollar system to watch us."[41]

It may be that anxiety over cameras is beginning to wear off as they become more widespread. One privacy expert, William Staples, says that in presentations he gives, older individuals still show anxiety about surveillance, while younger generations are becoming more accepting. He claims that from a generation that grew up in a world with Rodney King captured on video and in a culture that promotes voyeurism, immunity to cameras should be expected.[42]

Another example of people's growing comfort with cameras comes from Brian Curry, the chief executive and founder of EarthCam, a Web site that gets thousands of visits from people looking for webcams from around the world. EarthCam also operates it own cameras, which show video from Times Square. Curry claims that rather than being inhibited, people frequently are seen waving at the cameras. "We're offering a window on the world that is very much like sitting in a restaurant and looking out on the street. To try to inhibit this by saying it represents a brave new society where people are losing their privacy is far-fetched."[43]

The success of reality programs on television suggests that, rather than being intimidated by cameras, people are eager to expose their personal lives on prime time. Charles Sykes documents what he calls the exhibitionist society:

> [T]his orgy of pseudo-confession is so often regarded as testimony to the honesty and authenticity of the sharer. Under relentless battering from the therapeutic sensibility of our century, restraint and reticence—among the famous and obscure alike—are now regarded with something like distrust. Keeping one's dysfunction or family turmoil to oneself is not only regarded as suspicious but potentially unhealthy, whereas unburdening oneself—to a therapist, casual acquaintances, readers, or a national television audience—has come to be seen as a sign of healing. And even if the public occasionally expresses distaste for all of this, it still devours the details. Our appetite for self-revelation seems matched by an appetite for the chance to peep. As far as I can tell, no supermarket tabloid has gone out of business for lack of readership.[44]

Facial Recognition Doesn't Work

The most frequent criticism of facial recognition is that it isn't effective. Results of studies on the technology have so far been mixed, although recent research points to improvements in the last few years. For instance, in the use of facial recognition for one-to-one verification, where the goal is to confirm that a person is who they say they are, the technology is becoming exceptionally accurate, so much so that some systems can tell identical twins apart. Not quite as reliable is its use for one-to-many searches, that is, looking for a match among a large database of faces.

There is no denying that facial recognition technology is in the early stages of development and that there is room for improvement. It is well documented that factors such as lighting, eyeglasses, background objects, camera position, facial position, and expression can all impact the quality of the image, thus impacting the accuracy rate. In many of the most critical environments, such as at important security checkpoints, these factors can be controlled and aren't as much of a concern. Yet, if facial recognition is to become a widely used tool, it will have to demonstrate accuracy in a wide variety of situations.

In the meantime, as the technology of facial recognition improves, it can serve to supplement traditional measures of security. No one is arguing that a facial-recognition system should be the only factor to decide whether, for example, a person should be allowed on a plane. For instance, the University of Missouri, Rolla, which is using facial recognition to guard its nuclear reactor, has several other security measures in place. Anyone attempting to enter the research facility must pass through an occupied lobby, use a key at one door, and then use a keypad at another entrance.

Furthermore, precautions can be taken in the case of false positives (i.e., someone who isn't a suspect gets misidentified as one). The great thing about a facial-recognition system is that a human being can verify the results. After a hit is made, a security officer can take the flagged individual aside and do a careful comparison with the picture in the database. This is no different from when a police officer pulls over a suspect and compares his or her face to the image on a printed copy of a mug shot.

That fact that humans can verify the results illustrates that facial recognition and surveillance are nothing more than a technological extension of what police do today. John Woodward puts it this way: "To extend the analogy, a surveillance camera can be viewed as a form of mechanical police officer that watches or records events occurring in public places in which the person has no reasonable expectation of privacy."[45]

It's likely that many community police officers patrolling the beat will recognize some of the individuals they repeatedly see loitering in high-crime areas or cavorting with known criminals. With technology, what was once stored in the police officer's brain is now efficiently captured and digitized on a computer for other law enforcement officials to use.

Some Americans, especially minorities in the inner cities, who feel they might be targets of abuse, may even wish for cameras to replace some of the work currently performed by law enforcement officials. A video device can't misapply force or take away someone's liberties, and it may even record evidence when these actions occur. The reality is that there is much more to fear in this world than a camera on the street.

What's Your Frequency?

The introduction of new technology that can help identify and keep track of products as they move through the supply chain and into the consumer's hands has been surprisingly swift. One technology called Auto-ID uses small computer chips the size of a grain of sand that employ radio frequency ID tags (RFID) to communicate a unique serial number to a nearby scanner. With this technology, a pair of pants can communicate its location to a computer as it travels the supply chain, from its origination in the factory to a shelf in a retail store, dramatically reducing the chance that it will be lost or stolen and eliminating the need for manual tracking. The technology, which promises to revolutionize the tracking of inventory, was developed by the Massachusetts Institute of Technology (MIT) and has attracted some of the world's leading corporations and governments who plan to put RFID tags in everything from shoes to cash. It won't be long until it's possible for a refrigerator to report on its contents, a shopping cart to automatically bill the consumer's account, or a company to do a targeted recall of specific units of a bad product.

Although Auto-ID will help retailers recover some of the $70 billion dollars in lost inventory costs each year,[46] many are concerned that it will be used to keep tabs on consumers, for instance, recognizing when a person enters a store based on a RFID tag in his or her clothing or loyalty card. As in the case of facial recognition, this technology has parallels to the days when a customer entered a small-town general store and was recognized and greeted by the clerk. Is this really such a scary thing? If I go to Starbucks, and they have my favorite latte paid for and waiting for me when I enter, how much harm is done?

Of course, if Auto-IDs were used one day to track behaviors once a customer left the store, perhaps even in the home such as revealing how quickly a case of alcohol was finished, many people might become unsettled by the technology. Yet, there still could be occasions when a consumer may want to have his behavior in the home monitored, such as having the pharmacy alerted when an important prescription is ready to be refilled. The key is to ensure that greater openness enables consumers to know which products have radio-tracking technology and thus empowers them to decide for themselves in a free market whether to purchase that product or not.

A Network of a Million Eyes

It may turn out in the near future that concerns about surveillance were pointed in the wrong direction. With cameras seemingly following a Moore's Law pattern of growth, it may be that millions of watchful eyes from a public armed with cheap video recording devices will be a greater threat to privacy than any system the government could hope to implement.

In fact, U.S. citizens mobilized into a human network of watchfulness may be the most effective and far-reaching method of surveillance against the hidden plots of terrorists. Although cameras and other devices are slowly enveloping major cities in a web of surveillance, if there is ever to be the kind of coverage needed to identify suspicious activity in a country as large as the United States, an infrastructure already in place—human beings—will provide an option.

Using citizens to look for suspicious behavior and help solve crimes is nothing new in America. The use of mug shots, such as the FBI's Ten Most Wanted list, is an example. The listing, which was started in 1950, is extremely popular with the public and the Ten Most Wanted Web site is the FBI's most visited, with over 2.5 million hits per month.[47] After 9/11, the FBI added a Most Wanted Terrorists list with bin Laden headlining the group. The television program, "America's Most Wanted," takes mug shots and crime solving to another level by reenacting crimes in primetime. Hosted by John Walsh, the Web site claims that the program has helped solve over seven hundred crimes.

Local surveillance programs engage citizens on an even more active level. Many are familiar with Crime Stoppers and Neighborhood Watch. Crime Stoppers provides any person who has information on a crime with a toll-free number to call. Callers remain anonymous and can receive cash rewards if their tips lead to a successful prosecution. Neighborhood Watch programs, which have a long history in the United States, use local groups of volunteers, many of whom are senior citizens, to comb community streets looking for signs of unlawful activity.

Although Neighborhood Watch employs a more static infrastructure of citizens to look for trouble, the Amber Alert System can spring into action anywhere in the country at a moment's notice. Started in 1996 in response to the kidnapping and death of nine-year-old Amber Hagerman in Texas, the national program is a joint effort of law enforcement and broadcasters to alert the public to a child's abduction. In a well-known success story in 2002, two California teens were rescued from an abductor just minutes before their death when citizens provided tips to the police. Germany is experimenting with a similar program where volunteers in cities across the country sign up to have text messages sent to their mobile phones. These smart mobs could be activated at a moment's notice and used to help spot a criminal or terrorist on the run.

Well-established principles from the Neighborhood Watch and the Amber Alert programs could be applied in a national effort to watch for signs of terrorism. Citizens could be encouraged to call authorities if they see strange behavior, such as someone carrying materials into an apartment while wearing a chemical suit. Or, if there were specific intelligence about a terrorist threat, such as four men of African descent who had recently evaded authorities at a Canadian-American border crossing and were now driving a blue Chevy pickup truck through Washington State, the public could be alerted with the details.

In fact, the Bush administration did propose a terrorist watch program called the Terrorist Information and Prevention System (TIPS). The idea was short-lived as civil libertarians considered the program akin to spying, and the administration was forced to shelve it after a coordinated scare campaign. Unfortunately, few people were willing to stand up and argue that TIPS was little different from Crime Stoppers or the Amber Alert. Ironically, shortly after the program was disbanded, the entire country was transformed into a TIPS-like program in the search for the D.C. sniper.

Public surveillance will continue to cast an even wider net as digital recorders and cameras in cell phones and other handheld devices allow individuals to capture the world around them. We've seen this with the burgeoning popularity among teens of underground and bootleg DVDs containing reality recordings of everything from raucous parties to fist fights.

Various companies are developing technologies to facilitate the sharing of all this recorded information, either across the Internet or between electronic devices. It won't be long before someone can request footage of any location, for example, Dupont Circle in downtown D.C., to check on traffic or to ascertain whether parking is available.

Although this vision of millions of roving reporters with cameras might be scary to some, it's likely to be even more worrisome to the fanatics who will have to fret that anyone, including a neighbor, a friend, a coworker, or a complete stranger might uncover their devious plot. At the same time, many of these cameras will be shining their lenses back at public officials and those in power, supercharging the system of checks and balances in a way the Founding Fathers could have only dreamed.

CHAPTER 7

THERE'S GOLD IN THEM THAR DATA

The Case for Information Analysis

In combating an enemy that seeks to hide in the shadows and strike without warning, information becomes one of America's most important defenses. Unfortunately, as experts have pointed out, many of the strategies developed by U.S. intelligence for collecting information on cold war adversaries like the Soviet Union are no longer as effective in the antiterror campaign. For example, the familiar tactic of eavesdropping on an embassy is ineffectual against a group that doesn't have a country to call home. Using human intelligence to infiltrate a group like al Qaeda isn't a much better alternative when dealing with a leadership that puts a premium on relationships based on family ties and years of established trust.

It's not that information is impossible to collect on terrorist networks like al Qaeda, but that outdated strategies and tactics for intelligence gathering must adapt to this new threat. Even though these groups operate clandestinely, their members leave a signature behind in the form of electronic footprints whenever they purchase items, travel, use cell phones, or engage in any number of other transactions that are electronically recorded. In isolation, any one of these activities, such as taking flying lessons, will seem rather innocuous; when they are combined with other activities, like having a visa from a country that supports terrorism, calling a person on a terrorist watch list, or checking books on biological weapons out of a library, a larger pattern of potential terrorist activity can emerge.

Being able to see the terrorist forest despite the trees requires a complete rethinking of the role of information in intelligence operations. It includes broadening the range of sources from where data is pulled. It requires improved information sharing so that analysts from disparate agencies aren't like the proverbial blind men describing the white elephant. Lastly, it demands comprehensive analysis and data mining to enable patterns to emerge from a torrent of facts and details.

The Wall of Separation

Although the failures of the intelligence community prior to 9/11 have been widely publicized, I hope readers will indulge me as I summarize a few of the main findings because they are so relevant to the argument I will shortly make for IA.

The hearings conducted after 9/11 concluded, among other things, that it wasn't just problems with the collection of information that prevented a connecting of the dots, but with the sharing of it. Investigations revealed that facts on al Qaeda were known prior to the attacks; they just were not exchanged between agencies like the CIA, FBI, and INS.

The egregious case of Khalid al-Mihdhar and Nawaf al-Hamzi is a startling example of the state of information sharing in the intelligence community that existed prior to 9/11. In a much-referred-to account, the CIA received information from Malaysian intelligence in early 2000 that the two had been seen in Kuala Lumpur at a large gathering of al Qaeda operatives. Although the CIA knew the men had entered the United States, the agency never alerted the FBI or INS to their presence. After the bombing of the USS *Cole* in Yemen in October 2000, al-Mihdhar was seen in a picture standing next to the attack's planner, Khallad bin Atash; yet, the CIA failed to place him on a terrorist watch list. Al-Mihdhar's picture showed up again during a meeting between the FBI and CIA on July 11, 2001; however, FBI field agents were rebuffed when they pressed for details. According to the final report of the congressional joint inquiry into 9/11:

> When these photos were shown to us, we had information at the time that one of the suspects had actually traveled to the same region of the

world that this might have taken place, so we pressed the individuals there for more information regarding the meeting. . . .

And we were told that that information—as I recall, we were told that that information could not be passed and that they would try to do it in the days and weeks to come. That meeting—I wouldn't say it was very contentious, but we were not very happy, the New York agents at the time were not very happy that certain information couldn't be shared with us.[1]

It wasn't until August 23, 2001, that the CIA sent an urgent cable to the FBI and State Department notifying them about al-Mihdhar and al-Hamzi. By that point investigators would have precious little time to track the two down. Senator Richard Shelby (R-AL) offered a biting criticism of these events in his comments to the congressional joint inquiry into 9/11, saying, "Sadly, the CIA seems to have concluded that the maintenance of its information monopoly was more important than stopping terrorists from entering or operating within the United States."[2]

Breakdowns of this nature can be traced back to the long-standing U.S. policy of preventing the intermingling of intelligence and law enforcement information. For instance, Rule 6(e) of the Federal Rules of Criminal Procedure prevented data collected by the FBI during a grand jury investigation or intercepted from a court-authorized criminal wiretap from being shared with the CIA. In many instances, the CIA wasn't interested in information from the FBI unless it dealt with a particular subject, like espionage. This information barrier, the so-called wall of separation between the FBI and CIA, was the subject of countless hearings after 9/11 and blamed as one of the causes of the government's inability to connect the dots before the attacks.

For example, even though the FBI didn't learn about al-Mihdhar and al-Hamzi until just a few weeks before 9/11, the bureau still had time to pursue the two. However, an e-mail presented at one hearing revealed that because officials at FBI headquarters believed this to be an intelligence matter out of FBI jurisdiction, they hesitated to send FBI agents after the men. An e-mail from FBI headquarters to the agent citing a recommendation by the bureau's National Security Law Unit said, "This case, in its entirety, is based on [intelligence]. If at such time as information is developed indicating the existence of a substantial federal crime, that information will be passed over the wall according to the proper procedures and turned over for follow-up criminal investigation."[3]

The agent responded on August 29:

Someday someone will die—and wall or not—the public will not understand why we were not more effective and throwing every resource we had at certain "problems." Let's hope the [FBI's] National Security Law Unit will stand behind their decisions then, especially since the biggest threat to us now, [Osama bin Laden], is getting the most "protection."[4]

The bricks that form the wall of separation between law enforcement and intelligence were first laid at the beginning of the cold war with the creation of the 1947 *National Security Act*, which established the CIA. President Harry Truman was so concerned about the potential for domestic abuse by the CIA that he was determined to prevent it from participating in domestic law enforcement. The act says that the director of central intelligence shall "collect intelligence through human sources and by other appropriate means, except that the Agency shall have no police, subpoena, or law enforcement powers or internal security functions."[5]

The role of law enforcement was thus left to the jurisdiction of the FBI.

The wall was raised even higher in the 1970s when reports from the Church Committee and Rockefeller Commission revealed that some intermingling of law enforcement and intelligence activities had yielded negative consequences. One troubling example that came to light involved a program called COINTELPRO through which the FBI infiltrated and harassed political dissident groups for years.

As a result of the revelations of domestic spying, new restrictions were placed on information sharing, including guidelines issued by the Justice Department in 1976 to regulate foreign intelligence and domestic law enforcement investigations. In 1978 the Foreign Intelligence Surveillance Court (FISC) was created to regulate activities that involved collecting intelligence on a foreign power.

Over the years the fundamental differences in the structures and philosophies of the FBI and CIA have further contributed to the wall of separation, even though the CIA focuses on foreign-intelligence work and the FBI's raison d'etre is catching and prosecuting criminals. Unlike the United Kingdom, which has a separate domestic intelligence agency, MI5, the FBI combines domestic intelligence and law enforcement into one organization, with law enforcement getting the bulk of the attention and resources.

A disposition toward law enforcement and the gathering of evidence to be used in court encourages FBI agents to hoard information in order to protect it. In their minds, sharing details with the CIA might jeopardize the integrity of a criminal investigation or threaten individual privacy.

In addition, unlike their counterparts at the CIA, who focus on slowly and deliberately assessing intelligence in a strategic way, the FBI is more reactive, responding to specific criminal cases within shorter time frames. In fact, a majority of the 11,400 agents work on crimes that have already been committed.[6]

As a result of what Shelby calls a "casefile" mentality, analysts are not in a position to make connections to other cases in order to see the strategic big picture. For instance, prior to 9/11 there were numerous clues about a plot to use airplanes as a weapon against the United States. Isolated silos of information within the FBI included details on convicted terrorist Abdul Hakim Murad, who was part of a plot to blow up airlines over the Pacific Ocean; on the attempt of Libya to send terrorists to flight training school; on Zacarias Moussaoui, who had taken flight-school lessons in Norman, Oklahoma; and on Osama bin Laden's personal pilot, who had taken lessons at the same school two years earlier. The congressional joint inquiry into 9/11 indicated that intelligence sources had even eavesdropped on an operative who said, "plans to hijack U.S. aircraft proceeding well."[7] Even so, the agency did not piece the puzzle together.

On top of all those clues, the FBI had the Phoenix memo, practically a cheat sheet for connecting the 9/11 dots. In this July 10, 2001, electronic communication, Phoenix-based FBI agent Kenneth Williams warned that terrorists associated with Osama bin Laden might be attending U.S. flight schools. The memo began with a prophetic first sentence: "The purpose of this communication is to advise the bureau and New York of the possibility of a coordinated effort by Osama Bin Laden to send students to the United States to attend civilian aviation universities and colleges."

The memo, which also called for a nationwide sweep of flight schools in search of al Qaeda terrorists, had limited circulation within the FBI and never made it to the Federal Aviation Administration (FAA) until after the 9/11 attacks.

Other investigations have revealed that problems with information sharing aren't limited to the CIA and FBI, but are systemic, pervading the U.S. government. During one hearing, a State Department official testified that for more than a decade, the FBI refused the INS access to its National Crime Information Center Database on the grounds that the agency was involved in law enforcement.[8]

Improving Information Sharing

Since 9/11, some positive steps have been taken to remedy the stagnant flow of information between agencies. In particular, Sections 202 and 218 of the *USA PATRIOT Act*, which were designed to facilitate information sharing between law enforcement and intelligence agencies, appear to have paid dividends. The U.S. government had been watching Sami Al-Arian, a University of South Florida engineering professor, for years.[9] Wiretaps placed by intelligence agents revealed that Al-Arian was an international leader of the Palestinian Islamic Jihad (PIJ), one of the most violent groups in the Middle East. He was routinely heard consulting with terrorist leaders from PIJ and was known to have funneled money to the families of suicide bombers.

After 9/11, we shudder to think that an intelligence agency could sit for years on information about the existence of a likely terrorist conspirator on American soil without sharing it with other branches of government. This is a clear example of a well-intentioned policy that, when implemented, went too far and hamstrung efforts to protect national security. Fortunately, Sections like 202 and 218 in the *USA PATRIOT Act* effectively lowered the wall between the FBI and CIA and finally allowed authorities to arrest Al-Arian in 2003.

The merging of twenty-two agencies and 180,000 employees into the Department of Homeland Security (DHS) was another positive step forward in facilitating improved coordination and cooperation within the federal government. The DHS has become the primary institution responsible for pulling together information from intelligence, law enforcement, and public and private sources to better assess security threats. As the main integrator of data, DHS can serve as a single point of contact for state and local agencies and the private sector.

In another effort, the Bush administration created the Terrorist Threat Integration Center (TTIC). According to the administration, the TTIC's mission is to "close the 'seam' between analysis of foreign and domestic intelligence on terrorism."[10] It includes elements of the DHS, the FBI's Counterterrorism Division, the CIA's Counterterrorist Center, and the Department of Defense. It aims to be an "All Source" central clearinghouse for all intelligence information, including raw data and finished reports.

Critical to both the DHS and the TTIC will be working with local and state officials and private companies across the country. The challenge for homeland security is not in protecting a few metropolitan areas like New York City or Washington, D.C., but supporting local officials and first responders throughout the United States in their efforts to prevent and prepare for a possible attack. There may be over eleven thousand FBI agents working across the country, but that number pales in comparison to the hundreds of thousands of local and state government employees and private personnel that will play a critical role in responding to a threat.

To reach a level of integration throughout the United States will require significant investments in technology. Simply upgrading the FBI illustrates the difficulty of this challenge. After 9/11 it was reported that local FBI offices were using 56K modems and 486 processor computers and that many agents didn't have e-mail or Internet access. Six months of paper reports were waiting to be entered into the agency's Automated Case System (ACS),[11] and more than sixty-eight hundred leads dating back to 1995 were waiting to be assigned in the system.[12] The inefficiency of ACS was blamed for the temporary loss of more than four thousand documents related to the criminal prosecution of Timothy McVeigh, the convicted bomber of the Murrah Federal Building in Oklahoma City.

To remedy this technology deficit, the FBI is spending hundreds of millions of dollars on computer and technology upgrades in a project called Trilogy. Trilogy will provide for a common e-mail system, high-speed connections, and desktop computers that link field offices with FBI headquarters. Already beset with cost overruns, it remains to be seen whether the upgraded system will be able to go live in late 2004 as planned.

If getting the FBI databases to talk to each other is a challenge, imagine the hurdles involved in integrating eighteen thousand local, state, and federal agencies and thousands of private companies. Paramount will be the use of technology that emphasizes interoperability, such as common Internet protocols like TCP/IP and HTTP, virtual private networks, XML for database interconnectivity, and standardized commercial-off-the-shelf (COTS) software. Yet, connecting the information pipes between agencies can only go so far in the war against terrorism. The analysis of this information is the most critical component.

Information Analysis

Computer applications that mine large amounts of data for meaningful patterns have been in existence for years in many diverse fields. These IA applications help scientists predict complex weather patterns, enable companies to tailor marketing programs to the interests of their customers, and assist pharmaceutical researchers in developing new drugs and breakthrough treatments for diseases.

Applied to the antiterror campaign, IA would facilitate the examination of data from a variety of electronic haystacks in an effort to predict and prevent the next 9/11 attack. Using public and private sources of data that record the transactions necessary to be an active member of society, the system would allow analysts to piece together more efficiently and effectively any clues that a terrorist might leave behind.

In making use of this information, an IA program would not focus on creating a single, massive, centralized database; instead, it could use data from a variety of databases only at the time that it is needed and without removing it from its original source. In this manner, the technology could serve as the basis for a network that would tie together information across government agencies and greatly facilitate information sharing within the intelligence community.

Subject-Based Analysis

Once distinct sources of data can be linked in real time, a number of different IA approaches could be used to help analysts draw meaningful inferences about potential terrorist activity. One such approach, referred to as a subject-based analysis, would model how officials currently carry out law enforcement investigations. Typically, during the investigation of a crime the law enforcement officer will start with a "lead," perhaps someone who knew the victim, and work outward, looking for other associates who might have some involvement or otherwise have evidence. IA could help model this technique by sorting through large data sets to identify connections between people. In honing in on a suspect, IA could uncover other people in the person's personal network—the people to whom they make long-distance calls, friends they visit, and anyone with whom they exchange money—all valuable information, which investigators can utilize.

The attacks of 9/11 offer a good case study of how subject-based analysis could have allowed authorities to link and identify most, if not all, of the hijackers before 9/11. For instance, a starting point for analysis would have been two subjects, Khalid al-Mihdhar and Nawaf al-Hamzi, both of whom the CIA had identified as likely terrorists.

Tracking down the two hijackers would not have been difficult as they had blatantly used their real names in a long trail of transactions.

Al-Hamzi used $3,000 to open a checking account in his name at Bank of America. Al-Mihdhar bought a 1988 Toyota Corolla and had it registered in his name. Page 13 of the 2000–2001 Pacific Bell White Pages had a listing for "AL-HAMZI Nawaf M 6401 Mount Ada Rd. 858–279–5919."[13] According to the congressional joint inquiry into 9/11, FBI agents admitted that they were able to find al-Mihdhar's address on the Internet within hours after the 9/11 attacks.

Once al-Hamzi and al-Mihdhar had been located, IA could have been used to draw connections to the other seventeen hijackers. An examination of their places of residence could have determined that Salem al-Hamzi used the same address as Nawaf al-Hamzi or that Mohammed Atta and Marwan Al-Shehhi used the same address as Khalid al-Mihdhar.[14] Phone numbers for Mohammed Atta would have provided links to Fayez Ahmed, Mohand Alshehri, Wail Alshehri, Waleed Alshehri, and Abdulaziz Alomari. Majed Moqed used the same frequent flier number as Khalid al-Mihdhar. Ahmed Alghamdi and Majed Moqed opened a bank account with al-Hamzi and Hani Hanjour. Salem al-Hamzi and Abdulaziz Alomari opened their own bank accounts with the help of al-Hamzi.[15]

Considering that, once the CIA tipped them off to al-Hamzi and al-Mihdhar, the FBI only had a few weeks before 9/11, the likelihood of their using a traditional investigation successfully to link the other seventeen hijackers before 9/11 would have been extremely low. By contrast, an automated IA system may have made such an analysis possible within that short time frame.

Pattern-Based Analysis

A second approach to IA in the war against terrorism is what has been called pattern-based analysis. Unlike a subject-based approach, which starts with a suspect and works outward, a pattern-based investigation would start with a potential model of terrorist activity and work inward. Using existing knowledge about terrorist plans and behavior compiled by intelligence experts, detailed models could be used by IA to predict potential threats. As new intelligence filtered in, the models could be updated. So for instance, if eavesdropping on terrorist suspects revealed that a cell from the United Kingdom was planning to use rental trucks full of explosives to attack targets in the United States, IA could assist in finding patterns of behavior that might correlate with the threat. Information that fit into the model, such as individuals with visas from the United Kingdom who had recently arrived in the United States, rented trucks, and purchased bomb-making ingredients like fertilizer, would be flagged.

Other Forms of IA

As with many forms of technology, an IA program would serve as an extension of the work that humans already do when using information to draw inferences. In preventing or solving a crime, investigators carefully piece together a variety of related clues. Scientists gather evidence during experiments to test their hypotheses. Physicians assess a patient's symptoms to support a medical diagnosis. Technology like IA can work in a similar fashion, sorting through volumes of data, finding associations, and helping individuals draw certain conclusions.

In fact, IA might make the work of individuals more efficient by relieving some of the burden of picking through haystacks of raw data and allowing analysts to focus more on the most relevant details. The technology that has greatly expanded our ability to store data is quickly outpacing human capacity to sort through information.

Using data to identify patterns and draw inferences is not a new concept for either the public or private sector. Credit card companies use such programs, albeit much simpler ones, to identify fraud among consumer transactions. For example, if you make a large purchase in a state different from where you live, it's likely that a program will flag your account for suspicious activity, and you'll receive a phone call.

Information analysis already plays an important role in the public sector. The Transportation Safety Administration's (TSA) Computer-Assisted Passenger Prescreening System (CAPPS) is designed to increase airline safety by screening passengers. The Treasury Department has a program called the Financial Crimes Enforcement Network (FinCEN) that searches data for evidence of financial crimes like money laundering. The Centers for Disease Control (CDC) has an effort underway to analyze information from sources such as doctor reports, emergency room visits, and prescriptions with the hope that the system can rapidly detect an outbreak caused by a biological agent.

One short-lived government IA program was known as Terrorism Information Awareness (TIA). Partly as a result of the failure to connect the dots before 9/11, the Bush administration had pushed the Information Awareness Office (IAO) at the Defense Department's DARPA to research the feasibility of an IA program to fight terrorism. The TIA effort was charged with investigating new collaborative and analytical technologies to "put together the pieces of the puzzle."

Yet, before any meaningful research could be carried out by DARPA, privacy and civil liberty advocates descended on the program like a pack of wolves on fresh meat. Charging that TIA would enable the government to spy on citizens, advocates stirred up a froth of hysteria and fear in the media and on Capitol Hill that forced the Bush administration to close the IAO and Congress to end funding for TIA. Although DARPA jeopardized the program by appointing a controversial figure like Adm. John Poindexter of Iran-Contra fame to lead the effort and by failing to address forthrightly civil liberties concerns from the very beginning, it's unlikely that they could have ever appeased critics of TIA, who were determined to see the plug pulled at any cost. Going forward, it's likely that many of the ideas and much of the technology behind TIA will live on and be utilized by different agencies, albeit in a more piecemeal and discrete fashion than would have been the case with TIA.

Legal Limits on Information Analysis

Since 9/11, many have questioned whether the government has the legal authority to undertake a massive IA program in the war against terrorism. As of yet, it's not clear whether a traditional interpretation of the Fourth Amendment's restrictions on searches and seizures applies to IA. According to Stewart Baker, the Fourth Amendment "search model" does not fit the new IA techniques that are available.[16] For instance, the Fourth Amendment requires authorities to obtain a warrant for a search of an individual, but with IA, authorities may have to search and process large amounts of data before any particular individual is identified.

Courts have generally ruled that individuals do not have an expectation of privacy with information turned over to third parties. As a result, authorities can access many kinds of information with a simple subpoena instead of a warrant, including an individual's bank, telephone, and credit card records. Some information can be obtained even without a subpoena, including arrest records and driver's license information.

Compared to a warrant, which requires a judge to review evidence, a subpoena is a much easier means for authorities to obtain evidence. They may issue it on their own authority and, if challenged, they only have to show that the information is relevant to the investigation and that the request is not unduly burdensome. Baker suggests that such requirements place little burden on the government when it comes to gaining access to third-party data:

> The efforts to shoehorn new technologies into the "search framework" had its limits, however. One problem for privacy advocates was that the Fourth Amendment's privacy protection is personal—limited to the person who controls the "houses, papers, and effects." If police call on a suspect and want to search his house, they would need a warrant. But if they call on his mother, and want to search a suitcase he left with her, they can do so with her consent, not her son's. Similarly, if they call on his employer and want to search his work desk, they only need the employer's permission. What is more, even if the employer refuses to cooperate, a simple subpoena, not a search warrant, is usually sufficient to give the government access to things or information in the hands of a third party.[17]

Although the courts have not always provided strong protection for third-party data, Congress has stepped in to limit access to it by authorities. For instance, the *Federal Privacy Act* of 1974 was passed to prevent government agencies from sharing an individual's personal information with other agencies for a purpose other than for which it was originally intended. Some exceptions to the information-sharing restrictions are allowed, however, if the sharing is listed as a "routine use" and is published in the *Federal Register*.

The *Federal Privacy Act* was further strengthened in 1988 with the passage of the *Computer Matching Act*, which amended the former act to restrict the matching of individuals across different agency databases. One of the first such matching projects was conducted by the Department of Health, Education, and Welfare (HEW) in 1977 to identify federal employees who were illegally receiving Welfare benefits. The *Computer Matching Act* also took some of the teeth out of the *Federal Privacy Act* by excluding matches performed for foreign counterintelligence purposes.

Congress has passed a hodgepodge of laws that seek to regulate access to specific kinds of information. These include Title III (*Omnibus Crime Control and Safe Streets Act* of 1968) to protect electronic communications, the *Right to Financial Privacy Act* of 1978 to protect financial records, the *Cable Act* of 1984 to protect cable viewing records, the *Video Privacy Protection Act* of 1988 to protect video rental records, and the *Federal Education Records and Privacy Act* of 1974 to protect school records. These laws provide guidelines on accessing certain kinds of information and the penalties breaches of the rules. It is also true that many of these laws have exceptions for criminal investigations and/or national security concerns that expressly allow the government to subpoena records.

Restrictions on access to information have also been put in place through the *General Crimes Guidelines*, which were first instituted by Attorney General Edward Levi in 1976. These guidelines had restricted the FBI's access to the Internet, public records, and commercial databases prior to 9/11. These guidelines, which have now been broadened to allow for greater use of information, now say,

> The FBI is authorized to operate and participate in identification, tracking, and information systems for the purpose of identifying and locating terrorists, excluding or removing from the United States alien terrorists and alien supporters of terrorist activity as authorized by law, assessing and responding to terrorist risks and threats, or otherwise detecting, prosecuting, or preventing terrorist activities.

Another constitutional question for IA and behavioral profiling is whether such programs would pass the equal protection provided for by the Fifth and Fourteenth Amendments. Equal protection limits the ability of the government to rely on certain suspect factors like race, ethnicity, or national origin when taking an action. Unlike the Fourth Amendment, which applies only to searches and seizures, equal protection applies to all government conduct. So for instance, constitutional difficulties could arise if a government program identified a group of potential terrorist suspects on the basis of race alone.

In evaluating whether actions violate equal protection, courts rely on a strict scrutiny standard. To meet this standard, the government must show a necessary relationship to a compelling state interest. In the war against terrorism, national security interests would likely meet the standard.

Courts have also given leeway on equal-protection violations when (1) profiling factors such as race are informal and aren't part of a written policy, and (2) factors other than race or ethnicity used by authorities would have led to the same conclusion.[18] According to Eric Braverman and Daniel Ortiz, these exemptions provide enough latitude for an IA program to avoid any equal-protection problems.

> Although the government cannot use descent from any particular ethnic or racial group as a written profiling factor without proving that such a

factor was necessary to protect national security, a very high burden, it could use such factors as associating with a country known to harbor terrorists, visiting particular countries where terrorist groups are known to be and using banks or other institutions associated with terrorist groups without any difficulty. As a practical matter, equal protection will likely pose no great obstacle to effective law enforcement.[19]

In summary, it appears that the government has considerable latitude in using IA applications in specific cases of national security, although many questions remain to be answered. These include whether restrictions in the *Computer Matching Act* need to be amended to account for these new technologies and how the subpoena process would work for automated IA of private databases, especially when a suspect isn't immediately identified. Most important is the question of how legislation in Congress should be written to oversee IA and protect civil liberties without stripping away IA's fundamental purpose.

Safeguards

Any program that proposes to increase government power, especially one that involves the use of personal information, will have to address the fears of Americans over the protection of privacy and other civil liberties. Many worry that any new IA program could allow the government, in a pattern similar to what was seen with COINTELPRO, to build detailed profiles that could be used against political opponents and opposition groups or to deprive average citizens of constitutionally protected liberties.

In response to these concerns, a number of experts have proposed measures that would leverage oversight structures and safeguard mechanisms already in existence. Paul Rosenzweig of the Heritage Foundation has suggested a comprehensive program that would include the following:[20]

- Congressional approval and public debate before the program is implemented
- Creation of internal limits and guidelines that respect existing privacy laws
- Protection of individual anonymity by revealing identity only through the approval of a federal judge
- Approval by a Senate-confirmed official of each pattern query

- The use of pattern-based analysis only as a trigger for additional investigation
- Protections and legal remedies for false-positive identifications
- Congressional oversight and penalties for abuse
- Use of the technology solely for the purpose of terrorism investigations

Will most Americans be willing to accept an IA program as a tool against terrorists? If the appropriate protections are in place, identities are hidden, and profiles are not created, it is possible that some in the public will be amenable. Many people can do the mental calculus that says the probability that one individual out of millions of U.S. citizens ever being flagged as suspicious is nearly nil, while the chance the system might help the government foil a potential terrorist plot is likely to be higher. An infinitesimal risk for a more probable benefit might well be worth the trade-off.

Criticisms of Information Analysis

The Sturm and Drang raised by civil libertarians over TIA is a good starting point to address the criticisms of IA. Columnist William Safire gives an example of some of the attitudes toward TIA with an article entitled, "You Are a Suspect."[21] He is representative of those who suggested that TIA would be the ultimate tool to snoop on Americans. Unfortunately, the misconception that TIA would be used to create detailed profiles on Americans was perpetuated in many ill-informed articles in the press. In reality, TIA was to be used for temporary searches across multiple databases to look for suspicious patterns pointing to terrorist activity. As mentioned earlier, there would be no need for a centralized data warehouse, only access to multiple databases through a subpoena or court order and the use of temporary queries, the results of which would later be discarded. Only when a suspicious pattern turned up would an individual be identified, most likely after court approval was obtained.

Another criticism of systems like TIA is the concern with false positives (Type I errors), where the wrong person is flagged in the system and targeted as a terrorist. Some critics have charged that because there is no adequate identifier in America, John Smith, an innocent American, might get confused with John Smith, a wanted terrorist. One article from *Wired News* quotes IA expert Herb Edelstein, who was critical of TIA: "The data quality problem is enormous, but what's alarming is the danger of false positives based on incorrect data. Think of the number of people who get in trouble with the law because they have the same name as somebody else."[22] The story concludes by saying, "Despite widespread use of Social Security numbers in medical and financial records, there is still no 'unique identifier' that would allow the new system to track individuals with total accuracy."[23]

On the one hand, advocates like Edelstein fight against improving the national system of identification. Then they turn around and use problems with identification as a criticism against TIA. If the country implemented a secure Homeland ID system that guaranteed that every person's identity was unique, there would be far less chance of confusing individuals in a TIA type of program.

Although the problem of false positives is not unique to TIA, it should certainly be taken seriously. There may be times when an IA program specifically, or any security effort in the antiterror campaign generally, might make a Type I error and flag an innocent person as a suspect. In America, traditional law enforcement efforts give substantial protection against Type I errors, even at the cost of more Type II errors, otherwise known as false negatives, or failing to punish someone who is guilty. As Paul Rosenzweig notes, the maxim "it is better that 10 guilty go free than that 1 innocent be mistakenly punished" has a long history in the American legal system; however, as Rosenzweig also points out, 9/11 changes the calculus by significantly raising the cost of Type II errors. Whereas a Type I error at an airport may now mean a more intrusive search and interview for a passenger and possibly a missed flight, a Type II error could result in a terrorist boarding a flight and hijacking it to calamitous ends.

TIA was a direct result of the shift in philosophy to focus on reducing Type II errors. The U.S. government appears to have decided to place greater emphasis on preventing a catastrophic attack by looking for signs that might premeditate one. As Chapter 2 intends to demonstrate, the threat posed is significant enough that a policy of preemption has to be a major component of a counterterrorism program. Whereas in the past one might have argued that the United States could absorb the first punch from an enemy, today no country can afford to withstand the blow that terrorists can deliver, a fact that reshapes the entire view of security in the United States, including a possible tolerance of more Type I errors.

With that being said, there are many ways to keep the number of Type I errors to a minimum and resolve them when they arise with minimal collateral consequences. These include making sure that IA is only used as the basis for initiating an investigation, having humans take over once someone is flagged, notifying individuals when they are the subject of an investigation, putting the burden on the government to justify any deprivation of civil liberties, eliminating ill effects when a person is cleared, and providing individuals with a full range of due process protections, such as the right to appeal any decision to a judicial body.[24]

In the battle against the likes of al Qaeda, technologies of openness, such as Homeland IDs, surveillance, and IA give the U.S. government another set of arrows in its quiver. Unlike the movie *Minority Report*, where law enforcement used mind readers known as "pre-cogs" to arrest people before they commit a crime, these tools will be used to expose terrorists already in the process of planning their deadly attacks. These measures won't be foolproof, but they will significantly raise the bar for those trying to carry out another 9/11 attack.

Regardless of the number of safeguards put in place, there will always be a visceral response by the civil libertarian community over any attempt to grant the government additional powers. I think in most cases, these groups aren't questioning the intentions of government officials. They realize that public servants, like themselves, are for the most part doing their best to serve the country and make it a better place for their children. As history has shown, there have been times when the government has abused its power in order to spy on and harass a limited number of citizens and activist groups. These abuses have bred a deep suspicion and distrust among civil libertarians, who argue that power corrupts even the best-intentioned people.

The other story history tells, however, one that is less often mentioned, is the fact that each time the United States experiences an overreaching by the government, its system of checks and balances is activated, the light of public scrutiny shines, and changes are implemented into law to prevent abuses from reoccurring. For example, most people today would acknowledge that the wholesale internment during WWII of more than a hundred thousand individuals of Japanese descent, including seniors and children, many of whom were American citizens, was an egregious and regrettable act. Although the *Enemy Alien Act*, a law that is still on the books, authorized the internment, eight out of nine Supreme Court justices have renounced the act over the years.[25] As further proof of the country's regret, reparations were eventually paid to those who suffered.

One might ask if the country learned from this shameful experience or whether the same type of event could happen again today. I'd say that the attacks of 9/11 provided a good test case. In many ways, 9/11 was just as horrific as Pearl Harbor and, in the same way, led the United States to declare war on an enemy. However, instead of rounding up Arab Americans with U.S. citizenship, the government incarcerated fewer than eight hundred foreign visitors with immigration violations[26] and only detained three individuals under the enemy-combatant designation.

In fact, the FBI and other law enforcement agencies around the country have been reaching out to Muslim communities in an effort to jointly root out terrorists. A report on NPR described authorities in New Jersey who were going to great pains to learn about Muslim culture and to respect important customs, such as removing shoes before entering a mosque.[27] Many Muslims have been very receptive because they are proud to be Americans and don't want a few extremists becoming the symbol of what Muslims represent. This doesn't paint a picture of an Orwellian leviathan seeking to aggrandize power and restrict civil liberties at its every turn.

Throughout its history the United States has taken a path toward more freedom, accountability, and transparency. Although the country has at times gotten off track, the government eventually rights itself, and while it is not perfect by any measure, our government is more protective of civil liberties and freedoms than at any other time in our nation's history. Less than a hundred years ago, women couldn't vote, blacks were barred from many restaurants, and children worked long hours in deplorable conditions. Today there are more protections for more groups of people, such as women, minorities, seniors, disabled individuals, and homosexuals, than the Founding Fathers could ever have imagined.

Still, those who believe the government should be better equipped to fight today's modern threats do not think that because they have more freedom and liberty, Americans should trust government enough to give it greater powers without asking for anything in return. Proponents of a better-equipped government understand and acknowledge that power without accountability is a historical recipe for disaster. On the contrary, they want to explore the uses of surveillance, IA, and improved identification in the fight against terrorism, but only within circumscribed limits that demand transparency and oversight and lead to greater accountability.

Support for efforts like TIA also represents recognition by a growing number of individuals of the open society paradox—that security in the twenty-first century won't come through a crackdown on individual freedoms, but rather through greater transparency in how we live with those freedoms. Rosenzweig expresses the sentiment behind the open society paradox:

> One could conceivably adopt a purely preventative mode in responding to terrorist threats, enhancing security at airports, government buildings, and the like and relying on increased physical intrusions and identity cards as a means of forestalling the next attack. But if we are not to condemn ourselves to the "citadelization" of America, we must also consider a different tack—the use of predictive technologies to attempt to anticipate and thwart terrorist attacks before they occur. These technologies come at some potential costs to liberty, but with the very real prospects of gains in other forms of liberty. Absolute protection for electronic privacy necessarily leads to even less physical privacy.[28]

Although many Americans may be anxious for the country to win the war against terrorism, they will always be suspicious of government power and fearful of expanding it. As we will discuss later, if Americans are to allow technologies of openness to be directed at them, they will need assurance that the same technology is pointing back at their leaders.

PART III

Revisiting Privacy in an Age of Terror

In the twenty-first century, when rapidly developing technologies of openness will make it harder for terrorists to blend in with the crowd, we must assume that the glare of greater transparency will unveil elements of our own lives. Although reducing the threat of terrorism is a welcomed benefit, more openness stands in direct conflict with the country's long history of valuing privacy and respecting anonymity. It may be that in order to reconcile this conflict, Americans will be called upon to decide which elements of privacy must be protected at all costs and which might be conceded in the interest of greater security.

It's not often that the public is called upon to revisit its conception of privacy. Privacy is rarely brought into question; it is usually accepted as an all-American, apple-pie value, like life, liberty, and the pursuit of happiness. Everybody wants privacy for him- or herself and can rarely get enough of it. And because privacy seldom conflicts with other rights, a discussion about balancing it with other societal concerns occurs infrequently. Amitai Etzioni's book *The Limits of Privacy* (New York: Basic Books, 1999) is a rare example of a call to weigh privacy against other societal goods. For example, he believes that the right to privacy of sexual predators must be balanced with the community's need to protect itself from criminals who tend to display a high recidivism rate.

Getting to the core of what is most important about privacy will demand that each of us ask some difficult questions. There are many assumptions about privacy that we frequently take for granted but must examine in an honest debate about the issue. Although questioning some of these sacred cows may rankle some of the issue's most fervent supporters, it will allow the public to make a more informed choice when it comes to finding the appropriate balance between openness and privacy. Chapter 8 and several proceeding chapters begin this effort by presenting some of the more common views about privacy and inviting their examination.

CHAPTER 8

LIFE, LIBERTY, AND THE PURSUIT OF PRIVACY

The tension that one finds in the process of democracy, where the proper balance between competing interests is continually calibrated, is clearly evident in the homeland security debate. Those favoring stronger security measures seek to grant the government enhanced powers in the antiterror campaign; those concerned about guaranteeing civil liberties and protecting privacy fight to keep the government in check. Finding the appropriate equilibrium in difficult issues like these is never easy and is made even more challenging by uncompromising attitudes on both sides of the debate. We'll discuss one of these notions in this chapter, the idea that privacy is an absolute value on par with liberty that should never be limited.

A Brief History of Privacy

If the public were to reexamine its views on privacy, a first step would be to take a look at the history of the issue. Although most Americans cherish privacy, many would be surprised to learn that a modern right to privacy has been established in only the last few decades. In fact, privacy isn't even mentioned in the U.S. Constitution, which is somewhat surprising because many discussions of privacy treat it as a hallowed right along the lines of life, liberty, and the pursuit of happiness. To understand how the right to privacy has been established in such a short period of time, it is helpful to look at a condensed history of the modern privacy movement.

For the greater part of America's history, the legal framework for the idea of privacy was intertwined with the protection of property rights. With a foundation in English law, the courts broadly applied the concept of "a man's house is his castle" to cover a number of invasions against a person. For example, publishing the contents of a man's journal could be considered a violation of his personal property. Spreading slander about someone could be viewed in the same light because a man's reputation was considered his property.

The foundation for a more contemporary conception of privacy, and one disconnected from the boundaries of property rights, is usually traced back to an article coauthored in 1890 by Samuel D. Warren Jr. and future Supreme Court justice Louis D. Brandeis. The article, published in the *Harvard Law Review* by the two close friends, who went to Harvard Law School together, would become very influential in the debate over privacy rights throughout the 1900s, influencing hundreds of legal cases. According to Robert Ellis Smith, "In just 26 pages, the law review article succeeded in virtually creating a new right of action for aggrieved individuals where none had existed before."[1]

Brandeis and Warren argued that many past court decisions that had granted redress on the basis of private property ought to have been based on a principle of privacy. They believed that the concept of rights must evolve alongside advances in society and that the right to privacy was no different.

In their article, privacy was referred to as the "right to be left alone." Many have pointed out that Brandeis and Warren's campaign for privacy grew out of their concern with the intrusiveness of the press, particularly with the negative coverage in Boston papers of the Warren family's social life.[2] Their privacy crusade would continue when Brandeis, as a Supreme Court justice, advocated on behalf of the issue.

Although Brandeis and Warren provided somewhat of an intellectual foundation for the concept, the growth of government databases first drew the public's concern to privacy issues. The roots of the issue lay in the creation of the Social Security Administration in 1935 as part of President Franklin Roosevelt's plan for a government-regulated retirement program. Established by the *Social Security Act*, it required that a percentage of each worker's paycheck be deposited into a trust fund with a matching portion from the government. In order for the Social Security Administration to keep track of the millions of people in the program, they had to assign to workers what were called Social Security account numbers. Smith mentions that the need to "register" for SSNs created a political uproar, although within three weeks, most of the twenty-six million registration forms had been returned by workers.[3]

Over the next few decades, changes in how the SSN was used would have significant ramifications for the future of privacy. In 1943, Roosevelt issued an executive order directing agencies to use SSNs for tracking people and their accounts in order to reduce waste that would result from agencies creating their own numbering systems.[4] This didn't become a significant issue until 1961, when the introduction of computers motivated agencies like the Civil Service Commission and the Internal Revenue Service to use SSNs to track federal workers and tax returns, respectively.[5] Many consider this the beginning of a slippery slope as additional government agencies and private companies began using the number to keep track of people.

The expansion of the SSN coincided with the advent of computers, which enabled the government to store greater amounts of information on U.S. citizens. One project that stirred anxiety among many people was the National Data Center. A group of researchers in 1965 identified more than six hundred different government systems and recommended that the government adopt a centralized data system.[6] The system, to be called the National Data Center, was backed by the Bureau of the Budget (now the Office of Management and Budget) and was meant to streamline the data-management process of the federal government. Within a couple of years, as public awareness of the project grew, the idea lost support, and the proposal was discarded. For those concerned about privacy, the National Data Center represented the government's incipient Orwellian intentions and a harbinger of things to come.

During the 1960s and early 1970s, awareness of privacy issues began to percolate among the public. Much of the developing angst over privacy was directed toward the government as a string of abuses across the world were fresh in people's minds. Many had not forgotten the Nazi and Soviet regimes during and after WWII, which had showed the world the dangers of unchecked government power.

Revelations of abuses by the U.S. government did not ameliorate these concerns. During the Eisenhower administration, for instance, the FBI formed the Counterintelligence Program (COINTELPRO) to infiltrate political organizations, particularly within the Communist Party, and target their members for harassment and defamation. A 1963 order by President Kennedy authorized the IRS to turn over citizens' tax records to the House Un-American Activities Committee. The discovery that the FBI had conducted wiretaps of Eleanor Roosevelt, Martin Luther King Jr., Justice William O. Douglas, and others did nothing to quell public fears.

The Nixon administration proved itself to be no paragon of privacy protection when members illegally wiretapped more than a dozen former or current government officials and several journalists in an attempt to track down leaks. The Church Committee report of 1976 brought these and other abuses to light and served as an indictment of domestic intelligence gathering in the United States

Abuses by the government were not lost on individuals concerned about privacy. Vance Packard wrote *The Naked Society* in 1964, one of the first books to discuss threats to privacy, not only from the government, but also from private businesses.[7]

Alan Westin in 1967 wrote one of the classic works on privacy, *Privacy and Freedom*. Westin was in the vanguard of a nascent privacy movement when he encouraged people to think about privacy in terms of the use of personal information. Westin argued that the Constitution protected the privacy not only of a person's sayings, actions, and relations, but also of personal information. He defined privacy as "the claim of individuals, groups, or institutions to determine when, how, and to what extent information about them is communicated to others."[8]

Court Cases

Greater public interest in privacy captured the attention of the U.S. government as both legislative and judicial branches began taking action on behalf of privacy interests. One court case, *Katz v. United States* (1967), would have a significant impact on later privacy rulings.[9] During an investigation of Katz, a suspected gambler, the government had planted a listening device outside a public phone booth to record his conversations without his knowledge. The Supreme Court ruled that the Fourth Amendment, which protects against unreasonable search and seizure, applied to people and not just places and, therefore, should have been applied to Katz in this situation. More important, the court set a criterion for judging whether privacy is invaded by asking whether a person has an expectation of privacy and whether that expectation is reasonable according to societal standards.

As many people have pointed out, the *Katz* ruling was a double-edged sword for the protection of privacy. On the one hand, privacy protection was brought out of the home and applied to individuals wherever they might be. On the other hand, steady invasions of privacy have lowered people's expectations, leading the court to use this lowered standard to determine a privacy invasion, thus lowering people's expectations even further in what some have likened to a vicious cycle.

The real breakthrough for the modern privacy movement came in a series of reproductive cases when the Supreme Court laid the groundwork for a right to privacy in the U.S. Constitution. Most people accept *Griswold v. Connecticut* (1965) as the first case to establish a constitutional right to privacy. In this ruling, the Supreme Court overturned a Connecticut statute that prevented a married couple from using contraceptives, claiming that the law violated a married couple's right to privacy. After some fervid searching and creative wordsmithery, Justice William Douglas located the exact source of a constitutional right to privacy. He wrote, "specific guarantees in the Bill of Rights have penumbras, formed by emanations from those guarantees that give them life and substance."[10] Douglas was criticized by many individuals who were flummoxed by this unabashed creation "out of thin air, of a general and undefined right of privacy."[11]

A follow up case, *Eisenstadt v. Baird* (1972), confirmed and extended *Griswold* by overturning a ban on the distribution of contraceptives, even for single people.[12] Going one step further, the Court would permanently link privacy to reproductive rights in 1973 with the case of *Roe v. Wade*. The Court ruled that a ban on abortion was unconstitutional because it violated a woman's right to privacy.

Congressional Action

While the courts were finding protection for privacy in the Constitution, Congress was writing it into law. One of the first steps toward safeguarding privacy was the passage of the FOIA in 1967. Although its purpose was to give individuals the right to petition federal agencies and departments for certain kinds of information, the law included exceptions intended to protect the privacy of Americans.

Shortly thereafter, the *Fair Credit Reporting Act of 1970* was passed in response to the growth of credit bureaus and concern over the amount of personal data they were compiling. The act became one of the first efforts to regulate the use of personal information.

In 1973, a HEW commission organized during the Nixon administration published a very influential report on privacy. The HEW commission allowed the Nixon administration to look strong on privacy in the wake of the developing Watergate scandal. The HEW report recommended a *Code of Fair Information Practice*, something that privacy advocates have called a bill of rights for computer users.[13] The report has been extremely effective in shaping how people think about privacy, and its principles continue to influence the creation of privacy laws both in the United States and in Europe. The principles identify the right (1) to know if personal information is being or has been collected, (2) to know what information is collected and how it is used, (3) to prevent any use of the information without one's consent, (4) to correct any inaccurate or incomplete information, and (5) to hold the information holder accountable for the accuracy and security of the information.

In 1974, the *Federal Privacy Act* was passed partly in response to the Watergate scandal. It was designed to apply the principles behind the *Code of Fair Information Practice* to federal agencies. One of its requirements was that agencies list all their information databases, which exposed to the public the amount of personal data the government was collecting.

Other privacy actions taken by Congress during this time included the 1974 *Family Educational Rights and Privacy Act,* which restricted the information schools could collect and release on students; a 1974 Privacy Protection Study Commission, which examined privacy invasions of individuals by companies; and the 1976 *Tax Reform Act,* which restricted the IRS from revealing information on people's tax returns.

Privacy in the Digital Age

Leaping forward another decade or more, one finds the next great advance for the privacy movement arriving with the advent of the digital age. With the growth of computer networks beginning in the 1980s, the new privacy concern became the security of electronic communications. Congress took notice and passed the *Electronic Communications Privacy Act* (ECPA) of 1986 to apply the protections of the *Wiretap Act* to computer communications.

As the spread of electronic communications enabled companies to exchange data with suppliers, partners, customers, and investors, the demand for protecting this information began to grow. Development of encryption technology would provide an answer and eventually lead to one of the first significant showdowns between the government and privacy advocates in what became known as the Clipper Chip debate.

Encryption had very humble beginnings with roots in the 1976 proposal by two American cryptographers, Martin Hellman and Whitfield Diffie, of a type of encryption called a public-key system.[14] The system allows the sender and receiver to use a public and a private key to lock and unlock a message and keep its contents private. Reg Whitaker quotes an article in the *New York Times* that uses this analogy:

> Alice . . . has a safe deposit box that has two keys. One key, which Alice always keeps as a private key, can only open the box; the other key, which is copied and widely distributed as a public key, can only lock it. When Bob . . . wants to send a message to Alice, he grabs a public lock key, places his message in the open safety deposit box and then locks it. When Alice wants to read the message, she opens the box with her private key and leaves the box open. Thus only Alice can read, but anyone can send.[15]

The fruit from this research hit the market in the early 1990s in the form of a shareware program called Pretty Good Privacy (PGP), a free e-mail encryption software program developed by Phillip Zimmerman, which is still available on the Internet. (Visit www.pgp.com to try the freeware version). Although PGP was a hit with the online community, the U.S. government became concerned that criminals and terrorists would use the technology to hide their illicit activities. To counter this threat, the U.S. government offered its own encryption software known as the Clipper Chip to which it would hold a backdoor key. The idea sent the privacy community into a frenzy, and it organized against the proposal and argued that the Clipper Chip was a restriction of the freedom of speech.

As an alternative, the Clinton administration proposed a key-escrow system in which trusted third parties in the private sector would hold the keys. By 1998, however, there was obviously little public acceptance of the Clipper Chip, and plans for it were shelved.

At about the same time that the Clipper Chip controversy was heating up, the FBI began pushing for legislation to enhance its ability to conduct wiretaps. The bureau planned to pay the telecommunications industry to build into its networks technology that would allow the FBI to intercept communications. The FBI tried to assuage public fears by arguing that agents would still need a warrant approved by a judge before carrying out a wiretap. Despite negative publicity in the press and the dedicated efforts of privacy advocates, President Clinton signed the *Communications Assistance to Law Enforcement Act* into law in October 1994.

Privacy Groups

During these battles over privacy in the information age, privacy advocacy groups, such as the Electronic Privacy Information Center (EPIC) and the Electronic Frontier Foundation (EFF), came of age. Lawyers Marc Rotenberg, Dave Banisar, and David L. Sobel founded EPIC in 1994. The group gained notoriety in the early 1990s with a successful public relations campaign against Lexis-Nexis, forcing the company to shut down a database that published the SSNs of many individuals. EPIC also led an effort against Intel when that company announced its plan to use a serial number embedded in the Pentium III processor to identify individual computers.

The campaigns against Lexis-Nexis and Intel demonstrate that a small interest group like EPIC can have a significant impact on issues. EPIC's annual report for 2002 shows expenses of only $866,662. While the group's funds might be limited, however, its reach, especially across the Internet, is long. According to a *New York Times* article from 1999,

> With an annual budget of just $250,000 to cover salaries for three full-time lawyers and three half-time helpers, they don't even have a fax line. Yet with just the touch of the button, employees of the Electronic Privacy Information Center can reach more than 10,000 people, an international audience that is educated and technologically savvy.[16]

Another group that has established itself as a staunch defender of privacy, the EFF, was founded in 1990 by John Barlow, a former songwriter for the Grateful Dead; multimillionaire Mitchell Kapor, founder of the Lotus software company; and John Gilmore, one of the pioneers of Sun Microsystems. The group gained nationwide attention during several legal cases, including its defense of Steve Jackson Games.

Steve Jackson founded the role-playing game company, an employee of which was being investigated by the FBI for hacking. The FBI obtained a warrant to search the offices of Steve Jackson Games and seized computers and equipment from the company, almost putting it out of business. EFF helped Steve Jackson sue the government, and a court sided with Jackson, ruling that the FBI had violated the *Privacy Protection Act* and the *Electronic Communications Privacy Act*. Through the publicity surrounding this event, EFF garnered significant attention, particularly in the cybercommunity.

Defining Privacy

Although efforts by groups like EPIC and EFF have focused more attention on the issue of privacy over the last couple of decades, it seems as if the public's understanding of the concept is as murky as ever. Some of the blame lies with the U.S. government, which has used the Constitution, legislative efforts, and executive orders to stitch together a patchwork quilt of privacy protections that fail to adhere to any single guiding principle.

The nebulous meaning of privacy also arises from the efforts of privacy advocacy groups that have prevented a narrow definition of the term. According to one advocate, keeping privacy defined broadly is an important strategy:

> Privacy as a social value has defied many attempts at a definition, and that is good. A definition, especially in the legal field, is an exercise in power. It tends to limit the notion to the terms of the definition, whereas privacy should remain one of the last ramparts against the exercise of power on individuals.[17]

Although expanding the concept of privacy may seem like a calculated effort by some to broaden the base of support for the issue, it more likely reflects mere confusion about how to define the term. Since Brandeis and Warren suggested many years ago that privacy was "the right to be left alone," the concept has gone through more facelifts than many Hollywood stars. As mentioned earlier, Alan Westin saw privacy in terms of protecting information. Another scholar claims that privacy is the essential component in developing intimacy and interpersonal relationships.[18] Phillip Kurland defines three components of privacy: anonymity, secrecy and autonomy.[19] A litany of other writers has put their individual spin on the meaning of privacy, further strengthening its immunity to a single definition. Is it any wonder that the public's expectation of privacy has been lowered when confusion over the concept has them trying to figure out just what is reasonable?

Smith points to this expansion of the idea of privacy that began in the 1970s:

> Outside the courts, what was known as "privacy" was taking on new meaning. In an age of newly asserted individual and group rights, it came to include the right to control your own body and self, as well as the traditional "right to be let alone." Thus, "privacy" was used to justify defiance

of codes restricting types of dress and hairstyles—a growing area of conflict now that graduates of the Vietnam War protests and survivors of the permissive Sixties were reaching the workplace." Furthermore, in an age of computers and sophisticated surveillance devices, "privacy" came to include the right to know what information was kept on you in a databank and the right to correct that information. This notion of "informational privacy" also included an element from the traditional concept of privacy—a right of confidentiality.[20]

There seems to be no end to the number of offenses that can be recategorized as invasions of privacy. Jeffrey Rosen in *The Unwanted Gaze* attempts to redefine sexual harassment in this way.

> Recognizing unwanted gazes as an offense against privacy and dignity helps us to understand that other forms of sexual harassment commonly experienced by women may be better conceived of as invasions of privacy than as examples of gender discrimination.... In the 1970s, feminist film critics wrote about the objectifying quality of the "male gaze" in popular movies, which reduced woman, in their view, to eroticized body parts fit for ogling by male directors, cameramen, and spectators. These critics are correct, I think, when they describe the indignity that can result from being looked at in a way that substitutes a part of a woman's body for the whole of her personality, but that indignity is more precisely described not primarily as a form of gender discrimination but instead as an invasion of privacy.[21]

The association of privacy with human rights has also helped expand the term's meaning. Francis Fukuyama suggests that privacy rights have been caught up in the agenda of activists seeking to continually widen the net of so-called human rights.

> Over the past generation, the rights industry has grown faster than an Internet IPO in the late 1990s. In addition to the aforementioned animal, women's and children's rights, there are gay rights, the rights of the disabled and handicapped, indigenous people's rights, the right to life, the right to die, the rights of the accused, and victim's rights, as well as the famous right to periodic vacations that is laid out in the Universal Declaration of Human Rights. The U.S. Bill of Rights is reasonably clear in enumerating a certain set of basic rights to be enjoyed by all American citizens, but in 1971 the Supreme Court, in *Roe v. Wade*, manufactured a new right out of the whole cloth, based on Justice Douglas's finding of a right to abortion that was an "emanation" from the "penumbra" of the similarly shadowy right to privacy in the earlier *Griswold v. Connecticut* decision.[22]

Fukuyama is referring to the elevation of privacy to the status of a human right in 1948 by Article 12 of the United Nations' Universal Declaration of Human Rights, which states, "No one shall be subjected to arbitrary interference with his privacy, family, home or correspondence."[23]

EPIC released a report with Privacy International called *Privacy and Human Rights 2002*. The report starts off with the following statement: "Privacy is a fundamental human right. It underpins human dignity and other values such as freedom of association and freedom of speech. It has become one of the most important human rights of the modern age."[24]

Privacy as Liberty

In fact, as privacy expands to cover a whole range of assaults on the individual, many of the issue's most fervent supporters have begun to equate it with the idea of liberty. As privacy is redefined as a human right and put on par with something akin to freedom, it becomes viewed as an absolute and unbounded right that should never face limits. Amitai Etzioni has made this point, suggesting that the idea of privacy as an inalienable right is common among many privacy scholars and advocates. He quotes Jean Cohen, who said that "a constitutionally protected right to personal privacy is indispensable to any modern conception of freedom."[25]

The idea of absolute rights goes back several centuries to John Locke (1632–1704), whose theory of natural rights helped to define the principles of modern democracy. His *Second Treatise of Civil Government* proposed that all individual human beings originally lived in a state of nature where they possessed important rights under natural law, including life, liberty, and property. When people voluntarily come together to form a society, these inalienable, natural rights are to be protected by society. This line of thought greatly influenced the Founding Fathers, who made one modification to Locke's list of rights by changing "private property" to the "pursuit of happiness" in the Declaration of Independence.

Of course, most people know that in certain cases even the most important freedoms must be limited. For instance, the right of free speech and expression does not extend to sedition, slander, defamation, or obscenity. Rights can be restricted for any number of reasons, including protecting national security, preserving public order, or supporting the general welfare of society. Even human rights conventions, such as the European Convention on Human Rights, limit what they consider to be inalienable. It is generally agreed that only the convention's Article 2 on the right to life and Article 3 on prohibitions against torture are considered absolute.

Privacy is certainly an important component of a free society, one that most Americans cherish. However, there are times when it, too, must face limits. In many cases, courts have ruled in favor of restrictions to privacy. In *Vernonia v. Acton* (1995), for instance, the Supreme Court ruled that the privacy of student athletes can be limited in order to test for drugs.[26]

Conceiving of privacy as an unbounded right removes the possibility of balancing it with other social goods. Arguing from an absolutist position, those who champion the issue are forced into an all-or-nothing position that demands that privacy be protected at all costs and that trade-offs be avoided. These circumstances render compromise unlikely, especially in the current privacy/security debate.

The privacy-without-limits proposition also misses the fact that the issue has many dimensions. There is usually a continuum of concern on various privacy issues for Americans who recognize the big difference between unveiling their medical histories and the list of magazines to which they subscribe.

Perhaps the fundamental flaw in the view that equates privacy to liberty is that in most cases where there is a trade-off in privacy, it isn't freedom that is lost. A grocery store may track your every purchase and build up a detailed dossier on your food preferences, but that it no way limits your ability to buy its fruits or vegetables. Even privacy intrusions by the government, the institution that has the police power to threaten your freedom, do not necessarily equate to a loss of liberty. Surveillance cameras in downtown Washington, D.C., may remove the anonymity of tourists visiting the capital or of protestors picketing the White House, but they don't preclude those activities. The Supreme Court recognized this back in 1970 in the case of *Tatum v. Laird*, when it rejected the ACLU's claim that harm was done by the Army's having a surveillance database of anti-war protestors.[27]

Of course, if the government uses surveillance to harass people and prevent them from marching, the issue of privacy becomes secondary to concerns over the violation of more fundamental rights, such as freedom of speech or association. As Chapter 12 discusses, too often the focus is on so-called invasions of privacy when we should really be asking if more basic rights are being trampled upon.

Giving up the idea that privacy is equivalent to freedom and restoring it to its proper position as one right among many will not likely be popular among its advocates. Etzioni says,

> To reconceptualize privacy, a highly revered right, may seem offensive, almost sacrilegious. We traditionally view individual rights as strong moral claims with universal appeal, indeed we perceive them as inalienable rights. Although we also realize that individual rights were formulated under certain historical conditions, we tend to conceive of these formulations as truths rather than mores fashioned for a given time that are open to amendment as conditions change.[28]

In the antiterror campaign, the nation will be tasked with finding a balance between providing security for Americans and protecting important civil liberties like privacy. The belief that privacy cannot be limited under any circumstances short-circuits these efforts and makes reasonable attempts to strike a compromise virtually impossible.

CHAPTER 9

PRIVACY LOST

The plethora of books that proclaim its death imply that privacy has been slowly wasting away over time. It is as if there once was a Garden of Eden of privacy, an unspoiled world that was free of prying eyes where people went about their daily life hidden in a shroud of anonymity. Only after taking a bite out of the forbidden fruit of the technology age with its databases and surveillance cameras were people cast out of paradise into a world that exposed their nakedness. One privacy advocate admits the existence of this view:

> A further central, and more controversial, assumption in the privacy debate is that privacy is something that "we" once had, but now it is something that public and private organizations employing the latest information and communications technologies are denying us. This theme is represented in a large corpus of polemical literature in which Orwellian metaphors and imagery are prolific, even though "1984" came and went without any palpable change in the attention paid to privacy questions. Continually over the last thirty years publishers in North America, Britain and elsewhere have been attracted by this more polemical genre. The contexts may change, the technologies may evolve, but the message is essentially the same: Privacy is eroding, dying, vanishing, receding, and so on. Despite privacy laws, conventions, codes, oversight agencies and international agreements, some have argued that privacy is something of the past, to the extent that one prominent business leader could proclaim, in a much-quoted statement, "You have zero privacy anyway; get over it!"[1]

Privacy in Early America

Is there any truth to the idea that privacy has eroded over time? Historical accounts of early American life tend to challenge this notion. David Flaherty's *Privacy in Colonial New England,* an exhaustive study of life among Puritan settlers in New England, provides some insight into the state of privacy in early America.

The Pilgrims, who sailed from England on the *Mayflower* and landed at Plymouth Rock in 1620, established the Puritans in America. The primary purpose of their trip to the New World was to escape religious persecution by the Church of England. The Puritans, for whom religion was a cornerstone of life, believed in a rejection of worldly values and strict adherence to the principles of the Bible. Failure to adhere to inflexible rules of community behavior could result in punishment by the church and possibly an eternity of torture in Hell, according to fire-and-brimstone sermons given by Puritan ministers like Jonathan Edwards.

To ensure that members of the community were not tempted by the Devil, the church encouraged neighbors to be on the lookout for sinful behavior. The clergy encouraged people to spy and inform on those who violated community standards. As a result, very little in daily life escaped constant surveillance and prying eyes in Puritan communities. According to Flaherty, while privacy was desired, it was in conflict with the expectation of community scrutiny.

> Puritans were encouraged to subordinate privacy to the more pressing purpose of collaborating in the creation of a City Set upon a Hill for the edification of the rest of humanity. The implements of this communal spirit were a pervasive moralism, the concept of watchfulness, the encouragement of mutual surveillance, and the suppression of self to community goals. A search for privacy could be a threat to the spirit of the community, which was so strong in early generations in the New World. Towns should be established and run in a collective, cohesive, and communitarian fashion. Parents should carefully regulate behavior within their families for the suppression of evils and the advancement of Puritan ideals. The Puritan concept of the righteous life should be enacted into statutes, and these laws properly enforced, even if this enforcement required some lessening of personal privacy through the use of surveillance techniques.[2]

One method the Puritans used to enforce these ideals was the use of surveillance networks such as the town "nightwatch." Village residents who served in this capacity would wander around looking for people outside after dark, a prohibition in many towns. Watchmen on the lookout for disorderly or suspicious activity, such as drinking or skipping church, could tap into a network of informants who were paid to spy on neighbors and look for sinful behavior. Watchmen were even allowed to enter people's homes on occasion to look for evidence of transgressions. Not all of the watchmen took their job seriously though, and in typical Puritan fashion, towns had to hire watchmen to watch the watchmen.

Although nosy neighbors provided a web of surveillance over daily life, the clergy played the role of data collectors. The church managed a yearly census that recorded the names of all the town residents, their occupations, and other details about their lives. The clergy also kept track of births, deaths, marriages, and even salvations and heresies. The collection of data served many useful purposes, such as offering warnings of potential epidemics, providing information to relatives in England, and publishing employment statistics as a way to lure workers to the town.

Keeping track of the residents of Puritan villages was not inordinately difficult. Communities were composed of no more than a few thousand people, making it almost impossible to disappear anonymously into a crowd as modern Americans might do today. The places people went and the things they did were likely to be noticed by another resident of the town.

Unlike people today, who seek personal space by retreating to the privacy of their homes, most inhabitants of pre-Colonial times didn't have this luxury. Only the upper classes could afford a home of any size, and more often than not, families were forced to live together, sometimes with their servants.

It was not uncommon for families to open their homes to strangers, and occasionally visitors would enter homes without knocking. There were typically few beds in a house, and it wasn't uncommon for family members to sleep together, even with strangers, in the same bed. In fact, Puritans were so concerned about monitoring behavior that they passed laws prohibiting people from living alone.

Although homes were crammed on the inside, communities were often just as crowded on the outside. Many towns required that homes be built next to each other and located no more than half a mile outside of the town center. In order to find personal space, one might need to leave the town and head into the wilderness. At times, the only sure way of finding privacy may have been in one's thoughts. Flaherty states,

"A search for privacy, especially in the form of solitude, may have necessitated leaving the home. Absolute physical privacy was always available outdoors and during periods of darkness. Occasionally the colonists may have had to use psychological, rather than physical, means of withdrawal in order to be alone."[3]

Packing up and moving to another town to gain some anonymity was not always a desirable option. For one, travel was dangerous. Travelers had to face the wilderness, Indians, the weather, and wildlife. Once they made it to another town there was no guarantee that they would be allowed to stay. Anyone wanting to settle in a New England town was forced to register with the local authorities and then get the approval of its inhabitants.

Of course, Puritan influences waned over time, and the eventual western expansion of the United States gave people more space in which to spread out. As Smith indicates, however, churches and communities often followed each other west and maintained many of the same invasive practices.[4] For instance, although log cabins were a practical and popular choice of home and provided a secure defense on the outside, they provided little privacy on the inside. Those living under the same roof would eat, sleep, and occasionally walk around naked within the one large room that made up the cabin.

Privacy in the Industrial Age

Privacy in the early Industrial Revolution encountered many of its own challenges. As people moved out of the country and into the city to look for work in factories and mills, they often found cramped conditions completely lacking in privacy. In mill towns, workers frequently stayed in overcrowded boarding homes where several people shared rooms and everyone used a common kitchen and dining room.

In Lowell, Massachusetts, mills became notorious for the conditions under which the women they employed worked. Women toiled through long, fifteen-hour days under constant supervision only to retire to rooms shared with five or six others. Even if they had the energy to go into town to find some personal space, a curfew of 10:00 P.M. with doors locked made leaving an unlikely option.

As the population of America continued to disperse over wider areas, communications became ever more important. As is the case today, when people express concerns that their phones might be wiretapped or their e-mails intercepted, there were concerns over the privacy of early communication systems. For example, in the early days of the postal service, people regularly opened others' mail. Smith points out how the Founding Fathers in the new government fretted over the privacy of mail. He offers this quote from Thomas Jefferson: "The infidelities of the post office and the circumstances of the times are against my writing fully and freely. I know not which mortifies me most, that I should fear to write what I think or my country bears such a state of things."[5] It wasn't until the mid-1880s that adhesive envelopes were developed so that people could begin to feel more confident that the privacy of their letters would be protected.[6]

Although Puritans may have trampled on privacy for religious reasons, Smith suggests that Americans' inherent curiosity about the lives of others underlay threats to privacy for later generations of Americans. Foreigners who visited the colonies remarked on the extreme nosiness of Americans.[7] The early penny press newspapers in the United States owed their success to this American trait. Full of juicy details about people's lives, newspapers did not hesitate to jettison an individual's privacy to get a good story. Smith provides an example of one account: "Bridget McMunn got drunk and threw a pitcher at Mr. Ellis, of 53 Ludlow St. Bridget said she was the mother of 3 little orphans—God bless their dear souls—and if she went to prison they would choke to death for the want of something to eat. Committed."[8]

This kind of intrusive reporting of people's private affairs motivated Warren and Brandeis to write their influential article that inspired the privacy revolution. Westin claims that Warren and Brandeis were part of an upper stratum of society whose disdain for the press provided the motivation for their article. He says, "For the patricians, the gossip press, commercial advertising, and exposure of the doings of the socially prominent were aggressive and unjustified intrusions pandering to 'mass' curiosity and tastes."[9]

Although these examples suggest that privacy in early America could be hard to find, one might wonder whether people were even motivated by a concern about privacy. Could the desire for privacy be a social construction of modern American society? Research indicates that a desire for privacy isn't just a recent or cultural phenomenon. It suggests that many animals, including humans, have an innate need for personal space. Experiments with rodents have shown that crowded living conditions give rise to poor health, reduced life span, and death.[10] Other psychological research with humans, such as the famous Russian submarine experiment, has shown that a lack of personal space can be a major cause of stress.[11]

Although there appears to be some evidence for a universal desire for privacy, other studies undercut these findings by demonstrating that cultures define the boundaries of privacy very differently. One example comes from the Dobe Ju of East Africa, who do not use latrines and engage in sexual acts in front of their children, a practice greatly at odds with conceptions of privacy in the West.[12]

Misconceptions about Technology

One of the views driving the Privacy Lost mythology is the idea that technology has made privacy an endangered species. A common belief holds that any benefits from the information age that allow instantaneous communication around the globe are nullified when technology makes it easier to expose the contents of those same interactions. Simson Garfinkel, author of *Database Nation*, expresses this view:

> "Technology is neutral" is a comforting idea, but it's wrong. History is replete with the dehumanizing effects of technology. Although it's possible to use technology to protect or enhance privacy, the tendency of technological advances is to do the reverse. It is harder, and frequently more expensive, to build devices and construct services that protect people's privacy than to destroy it.[13]

The disdain that many privacy advocates have for technology is reminiscent of another well-known movement, that of the Luddites. The Luddites were a group that formed in England in the early 1800s in response to the increased use of machines like the power loom, which put large numbers of textile workers out of work.

What began as a collection of anonymous letters sent to mill owners, each signed by a mythical Ned Ludd, quickly turned into a violent campaign. Unable to feed their families, workers went on a rampage across England, attacking and destroying factories and forcing Parliament to bring in troops to put down the uprising. Eventually Parliament passed the *Frame Breaking Act* in 1812, which allowed people convicted of machine breaking to be sentenced to death. By 1813, the movement disappeared almost as quickly as it began.[14]

As with the Luddites, modern-day criticisms of technology are frequently misdirected. In most cases, it is not the technology that should draw one's ire. Airplanes can be a wonderful transportation technology, or they can cause destruction as a deadly missile. Surveillance cameras can be used to catch a lawbreaker during the commission of a crime or to peep at women. Rather, the issue should be how people choose to use technology and what kinds of safeguards and penalties are put in place to discourage abuse.

What's missed in the privacy debate is that, in many ways, technology has given people in the modern age more privacy. Toby Lester, writing in the *Atlantic Monthly*, made this point:

> One of the earliest technologies, writing, enabled a new and enduring form of private communication. The printing press popularized reading, an intensely private affair. The wristwatch privatized time. Cheap and widely available mirrors allowed, literally, a new level of private self-reflection. The gummed envelope boosted expectations of privacy in the mail. The technological advances of the Industrial Revolution led to the creation of a prosperous middle class that could afford to build houses with separate rooms for family members. The single-party telephone line allowed for direct, immediate, and private communication at a distance. Modern roads and mass-produced automobiles made private travel possible. Television and radio brought news and entertainment into private homes.[15]

Privacy-Protecting Technologies

Paralleling the rise in privacy concerns has been the creation of a whole host of companies offering privacy-enhancing technologies. Are you concerned about how much information is revealed by your browser when you surf the Internet? Visit anonymizer.com to get an electronic privacy profile. I recently did an analysis of my Internet browser, which although interesting, didn't reveal too much. It identified my IP address, my type of browser, my version of windows, and whether I'm accepting cookies—all pretty innocuous stuff.

Interested in protecting the contents of your e-mail? Some companies offer a Web based e-mail system like Hotmail, but with 2,048-bit encryption. Other companies offer e-mail that will automatically self-destruct like a *Mission Impossible* missive.

Want to send e-mail anonymously? There is software that will remove the return address and IP address of your e-mail and make its origin very difficult to trace.

Concerned that someone is monitoring your keystrokes through Trojan horses and other computer-snooping programs that can be planted on a computer to record what you type? Products are available that plug in to your keyboard port and encrypt all of your keystrokes.

Worried that your ISP is tracking the Web sites you visit? Many companies offer software that will encrypt the Web page requests you make in your browser and then send the requests through their servers. Your ISP will see garbled Web addresses, and the site you are visiting will see its own domain address instead of yours.

Want to protect the files on your computer? There is software that secures your computer's files with encryption. Other software performs the same function for personal digital assistants like the Palm Pilot.

Many of these products rely on readily available encryption that offers users incredible levels of privacy and security. Simon Singh, author of *The Code Book: The Science of Secrecy from Ancient Egypt to Quantum Cryptography*, writes, "I could send you a message encrypted through free software on the Net, and the combined forces of the GCHQ, the NSA, the CIA and the FBI wouldn't be able to crack that code. And if they did manage to," he added, "I could just re-encrypt with Version 2.0."[16]

In another version of the arms race analogy, when a technology that threatens privacy is identified, a company looking to make a profit frequently comes up with a solution. Consider the spyware phenomenon. This is software designed to track an Internet user's activities and send the information to companies who sell the data for marketing purposes. The devilish aspect of spyware is that is bundled into other applications and surreptitiously installed on a computer without a user's knowledge.

In a rush to help Internet users out of these dire straits, numerous companies have developed products to battle spyware. I personally use a program called SpyBot, which scans my hard drive for malicious cookies and hidden software. I get online updates from the company that develops it in order to keep the program armed against the latest spyware tactics.

In light of a self-interested market economy driving the development of products that safeguard individual privacy, Scott McNealy of Sun Microsystems jokes that many of our worst fears are misplaced. "You can go and find a mailbox right now, open the door to a tin box, tin door, no lock, with unencrypted information in English, sealed in a paper-thin envelope with spit, yet people are worried about online privacy."[17]

This is not to suggest some idealized view of technology as an elixir that can cure all the privacy ills of the modern world. That view would be as misguided as the mythological view that there once existed a paradise of privacy. As we'll see in later chapters, the flow of information is becoming ever more difficult to stem, regardless of the number of privacy-enhancing technologies available. Consumers may not always be willing to take the time to utilize privacy tools, instead choosing to trade information for conveniences, such as online customized services.

However, we are advised to suspend the rush to judge new technologies as threats to privacy. Counterterrorism operations will require a host of new technological developments to uncover the hidden activities of groups like al Qaeda; however, just as we saw with TIA, rampant speculation about a technology's potential to obliterate privacy before it has advanced beyond the conceptual stage, premature verdicts, fear mongering, and calls for prohibition may leave us without tools against an enemy who won't hesitate to use technology against us.

CHAPTER 10

BIG BROTHER IS WATCHING YOU

O f all the so-called threats to privacy, none seem as ominous as Big Brother. It is hard to find a story on privacy these days without some mention of Big Brother lurking in the shadows, ready to seize someone's civil liberties at a moment's notice. Many of these references to Big Brother speculate about a future when the U.S. government is transformed into a repressive, totalitarian regime. Simson Garfinkel provides one vision of Big Brother:

> We can easily picture a society (our century has also done this for us), in which the police have the right to burst into any home, or any room, at any time; in other words, a society in which there was no place that is off-limits, no place where we are safe and where we could hide. Or a society that restricted personal choices, regulating whom you can marry, how you can raise your children, or what you can read in the privacy of your own home.[1]

Frequently, Big Brother is mentioned in the same breath as Hitler or Stalin to infer that one day our government might come to resemble those heinous dictatorships. One such view belongs to privacy advocate Phyllis Schafly, who claims,

> Two of the principal mechanisms by which the rulers of 20th century police states maintained their control over their people were the file and the internal passport. . . . These two methods of personal surveillance—

efficient watchdogs that prevented any emergence of freedom—required an army of bureaucrats fortified by a Gestapo, a Stasi or a KGB, plus the ability to commandeer an unlimited supply of paper and file folders. Technology has now made the task of building personal files on every citizen, and tracking our actions and movements, just as easy as logging onto the Internet.

Sometimes even politicians in the U.S. government get into the act. Representative Steve Chabot of Ohio once called a proposal for a computerized worker registry "1-800-BIG BROTHER."[2]

Orwell's Big Brother

Of course, we have George Orwell to thank for the Big Brother metaphor. Orwell's book *1984* tells the story of a totalitarian government, ruled by The Party, which uses complete control of the population to stay in power. Big Brother symbolizes The Party, and his face with its piercing stare is plastered on posters everywhere with the slogan "Big Brother Is Watching You." The Thought Police enforce the rules of Big Brother and monitor what people say and do through telescreens, large video devices that allow The Party to display propaganda while monitoring citizens.

Although *1984* is a work of brilliant imagination and superb storytelling, the fact is that Orwell got many things wrong. Orwell was worried that technology paired with totalitarianism would make governments like the Soviet Union an unmatchable menace that would threaten liberty far into the future. By 1948, when the book was written, Orwell had witnessed how technology had enabled Hitler to spread Nazi propaganda and Stalin to take over numerous satellite countries in Eastern Europe. By the year 1984, many commentators writing about the book claimed that Orwell's vision of totalitarianism was still on track.

However, with the fall of the Berlin Wall in 1989 and the collapse of the Soviet Union in 1991, it became clear that not even technology could prop up an economically bankrupt regime like the Soviet Union. In fact, technologies like the fax machine, copier, and short-wave radio allowed people inside the Soviet Union, and China more recently, to communicate with the outside world in order to undermine their leaders' control. One such example is described in a biography of George Soros, the billionaire who used his wealth in the 1980s to donate hundreds of copier machines to his native country of Hungary to help facilitate the dissemination of information during the Soviet's repressive rule.[3] Soros is extending his mission for greater openness to other countries around the globe through the work of a variety of foundations, including the Open Society Institute.[4]

In the past, dictators and brutal regimes relied on propaganda and tightly controlled messages to keep their subjects in check. With the technology of the information age, leaders are finding it much harder to prevent information from flowing across their borders. For example, the Iranian government is struggling to regulate the explosive growth of Internet cafés, which provide Iranians with news from around the world and an alternative perspective to the government's ideology.[5]

The same is true in China, where an army of online government censors attempts to block information deemed inappropriate. The CIA is trying to make this more difficult by investing in U.S. companies that provide technologies that disguise the source and content of information downloaded in China.[6] The Voice of America, an international broadcast that presents an American point of view abroad, is also helping to break down barriers in the Communist country by sending daily e-mails in Chinese to over 180,000 people.[7]

A more recent example occurred during the 2003 Iraq War when the United States targeted the e-mail and cell phones of officials in the Iraqi military and dropped hundreds of thousands of leaflets from the air in an effort to undermine confidence in Saddam Hussein and encourage surrender. American Air National Guard communication planes circling high above Iraqi territory assured the broadcast of coalition messages over standard AM/FM radio, television, short-wave, and military communication bands. One wonders if the evil dictators from the past like Stalin and Hitler would have lasted as long as they did had the citizens of the country had a pipeline of information from the free world.

The U.S. Government as Big Brother?

Do references to the U.S. government as a Big Brother represent reasonable fears? Although the future is unpredictable, it is somewhat of a stretch to suggest that the world's oldest and most open democracy is even close to becoming the fiction imagined by Orwell. One might argue that there have been times, such as during war, when the government has put limits on certain liberties. For example, there were the *Alien and Sedition Acts* of 1798, which were used to suppress criticism of the government during the Adams administration. In another frequently cited example, Abraham Lincoln suspended habeas corpus during the Civil War. In 1917, more than two thousand people were prosecuted under the *Espionage Act* for opposition to World War I. In World War II, Roosevelt authorized the internment of more than one hundred thousand Japanese-Americans.

In these historical references to the U.S. government as Big Brother, many fail to acknowledge that the system of checks and balances put in place by the Founding Fathers has worked time and time again, limiting oversteps in power and enabling corrective action whenever abuses have occurred. For example, in 1866, one year after the end of the Civil War, the Supreme Court ruled Lincoln's suspension of the writ of habeas corpus unconstitutional.

Many additional checks and balances involve oversight and transparency measures implemented in the last century. For example, the FOIA was passed in 1958 and again in 1966 in order to "open agency action to the light of public scrutiny."[8] The *Government in the Sunshine Act* of 1976 also requires many federal agencies to open most of their meetings to the public under conditions similar to the FOIA. Other government accountability measures include three bills passed in 1978: (1) the *Inspector General Act*, which established inspectors general to use full subpoena and investigation authority to root out fraud and waste in agencies; (2) the *Ethics in Government Act*, which forced top officials of all three branches of government to disclose financial statements and imposed restrictions on former employees lobbying their former agencies from the private sector; and (3) the *Presidential Records Act*, which required that most presidential papers become public one a president leaves office.

In addition, one might ask, if the federal government is such a threat to privacy, why has it been so busy passing legislation to protect privacy over the past three decades. Although many of these laws have been enacted to protect privacy in the private sector, their passage demonstrates the concern that lawmakers have for individual privacy. Some of the bills include the following:

- *Fair Credit Reporting Act* (1970)
- *Family Educational Rights and Privacy Act* (1974)
- *Privacy Act* (1974)
- *Tax Reform Act* (1976)
- *The Right to Financial Privacy Act* (1978)
- *Cable Communications Policy Act* (1984)
- *Cable Privacy Protection Act* (1984)

- *Electronic Communications Privacy Act* (1986)
- *Computer Matching and Privacy Protection Act* (1988)
- *Video Privacy Protection Act* (1988)
- *Telephone Consumer Protection Act* (1991)
- *Driver's Privacy Protection Act* (1994)
- *Children's Online Privacy Protection Act* (1998)

In recent years, privacy has continued to garner significant attention from Congress. Privacy scholar Fred Cate mentions that in the 104th Congress, almost 1,000 out of 7,945 bills addressed some privacy issue.[9] Ari Schwartz, a policy analyst for the Center for Democracy and Technology, claimed that the 106th Congress was "the most privacy focused, ever" and that "they held more hearings on privacy and on a greater range of issues than any before them."[10]

In many cases, Congress reacts very quickly to pressing privacy needs. When it was revealed during the nomination hearings of Robert Bork to the U.S. Supreme Court that his video-rental records had been given to a reporter, Congress moved apace to pass the *Video Privacy Protection Act* of 1988. At the state level, Florida passed the *Family Protection Act* in response to public outcry over attempts to reveal Dale Earnhardt's autopsy pictures after his 2001 death in a racing accident.

The USA PATRIOT Act

Since the fight against terrorism began, a steady drumbeat of diatribes and innuendo has suggested that the U.S. government is trying to roll back protections for civil liberties. The impetus for this criticism came on October 26, 2001, when the president signed the *USA PATRIOT Act* (*Uniting and Strengthening America by Providing Appropriate Tools Required to Intercept and Obstruct Terrorism*) into law, one day after the act passed Congress with only one vote in the Senate and sixty-six in the House dissenting.

The act gave the government a variety of increased powers in the antiterror campaign, including expanded wiretapping and surveillance authority, improved ability to share foreign-intelligence information, increased funding for border control activities, broader powers for the INS to deport noncitizens associated with terrorist organizations, and the ability to hold terrorism suspects in detention indefinitely.

The *USA PATRIOT Act* has many in the civil liberties community in an uproar of cosmic proportions. One ad campaign launched by the

ACLU in 2002 claimed that the Bush administration was using the act to violate civil liberties, including kicking people off airplanes because of the color of their skin. One commercial shows someone who represents Attorney General John Ashcroft cutting up the Constitution, crossing out sections, and writing in his own words. The voice-over says, "He [John Ashcroft] sees powers for the Bush Administration . . . the right to intrude on your privacy."

Domestic Surveillance

Civil libertarians point to the *USA PATRIOT Act*'s enhancement of the government's domestic surveillance authority as an example of the abuse of power. As Chapter 7 shows, for many years the FBI has been limited in its ability to participate in the domestic spying business. One such restriction comes from guidelines originally set up by the Justice Department in 1976 to address abuses by the FBI in infiltrating activist groups. FISC also put some limits on the FBI's ability to carry out domestic surveillance. This court was established by the *Foreign Intelligence Surveillance Act* of 1978, which imposed a formal process for obtaining wiretaps in national security investigations.

One of the loudest complaints about the *USA PATRIOT Act* is that many of these restrictions on surveillance have been circumvented. In the past if the FBI wanted to conduct surveillance, such as using a wiretap, for law enforcement purposes, it had to go through a court to get a warrant where the standard required a showing of probable cause that the person in question had committed a crime. The other option, going to the FISC, required that there be probable cause that the suspect in question was an agent of a foreign power and a showing that "the purpose" of the investigation was foreign-intelligence gathering, a phrase that the court had interpreted to mean "the primary purpose." The concern is that the *USA PATRIOT Act* in Section 218 lowered the standard for the latter, changing the requirement for foreign-intelligence gathering to be a "significant purpose" of the investigation, making it much easier for the FBI to carry out FISA surveillance.

Lost in the furor over FISA-approved surveillance orders is the point that the use of this provision against the majority of Americans is highly unlikely. FISA stipulates that for "U.S. persons" there must be a reason to believe that

> the target of the electronic surveillance is a foreign power or an agent of a foreign power: Provided, That no United States person may be consid-

ered a foreign power or an agent of a foreign power solely upon the basis of activities protected by the first amendment to the Constitution of the United States.[11]

The attorney general must make a showing of probable cause for the above to the FISC before an order for a wiretap is approved. There is little reason to believe this standard would apply to average Americans in more than the most infinitesimal number of cases. In fact, the number of FISA wiretaps and searches granted in 2002, a mere 1,228, supports this claim.[12] Although the 2002 numbers represent an increase from the 934 warrants issued in 2001, considering the size of the population and the number of foreign visitors to the country each year, this number is still extremely small and stands in stark contrast to the amount of press coverage that has been engendered. At the same time, traditional wiretap applications presented to a "normal" federal court decreased by 9 percent from 1,491 in 2001 to 1,359 in 2002. For groups like the ACLU to portray the government as interested in spying on average citizens and the *USA PATRIOT Act* as a free pass to surveillance is to foist a fiction on the public that foments an unnecessary level of hysteria.

Although many critics have been busy suggesting that the change of standard will allow the government to spy on average Americans, few have been willing to admit that restrictions on domestic surveillance, such as the wall of separation discussed in Chapter 7, played a role in the government's inability to prevent the 9/11 attacks. We've already seen how issues with FISA discouraged the FBI from tracking down Khalid al-Mihdhar and Nawaf al-Hamzi after the CIA warned the bureau about them just weeks prior to 9/11. The case of Zacarias Moussaoui provides another example of the past inadequacy of the FISA process.

In August of 2001, the officials of a flight school in Minneapolis had tipped the FBI about suspicious behavior by Moussaoui. Officials reported that Moussaoui had paid in cash to take lessons on a Boeing 747 jumbo jet and wanted to learn to fly the plane, but not how to take off in or land it. After the FBI's Minneapolis office detained Moussaoui on a visa violation, agents immediately determined that he had terrorist links and sought a warrant to search his computer.

In a decision that would have serious ramifications, FBI headquarters rejected the Minneapolis field office's attempts to obtain a criminal warrant on the grounds that doing so might be an admission that this was a criminal case, thus eliminating its chances of going through FISA. When agents pursued the FISA route, they ran into another roadblock as FBI officials in Washington, D.C., decided that Moussaoui, who had ties to Chechen terrorist organizations, was not connected to an established foreign power. Although it was based on bad legal advice about what constituted a foreign power, some have suggested that this decision would not have been made had FISA not been so critical of past FBI applications, something that had a "chilling effect" on the process.[13] The congressional joint inquiry into 9/11 concluded,

> During the summer of 2001, when the Intelligence Community was bracing for an imminent al Qaeda attack, difficulties with FBI applications of *Foreign Intelligence Surveillance Act* (FISA) surveillance and the FISA process led to a diminished level of coverage of suspected al Qaeda operatives in the United States. The effect of these difficulties was compounded by the perception that spread among FBI personnel at Headquarters and the field offices that the FISA process was lengthy and fraught with peril.[14]

According to Senator John Edwards of North Carolina, if agents had succeeded in getting the warrant from FISA, it would have led them to other al Qaeda members, including hijackers Khalid al-Mihdhar and Nawaf al-Hamzi.[15] One of the agents had even made the statement prior to 9/11 that he hoped Moussaoui "did not take control of a plane and fly it into the World Trade Center."[16]

Angry Librarians

Another measure in the *USA PATRIOT Act*, Section 215, often referred to as the "angry librarians" provision, has attracted an inordinate amount of criticism. Civil libertarian and librarian groups have been up in arms over the *USA PATRIOT Act*'s expansion of the kinds of items that the FBI can now request during intelligence gathering, including library records. Some librarians have even threatened to destroy borrowing records lest they be turned over to federal authorities.

All the rhetoric overlooks the fact that this section of the *USA PATRIOT Act* extends an authority that already exists in law enforcement investigations. Records have been subpoenaed from public libraries in a number of high-profile criminal cases, including the Versace murder case and the Unabomber investigation. Considering the evidence showing that the 9/11 terrorists used Internet connections at libraries, it would be natural for authorities to want to look at library records in certain cases. Of course, obtaining these records cannot be done willy-nilly; nor are these efforts likely to be targeted against average Americans. Investigators must get a court order from the FISA court. The process is different from that for standard grand jury subpoenas, which are used in criminal cases and which are requested without court supervision and contested only after they are granted.

Sneak and Peek

"Sneak and peek" is the name given to Section 213, which allows authorities to delay notification of the subject of a search warrant. Such an authority can be useful in cases where investigators believe a suspect might escape or destroy evidence. Although many individuals were surprised to learn of this practice after 9/11, this authority actually has a long history in crime fighting and has been upheld time and again by the courts. For example, if law enforcement authorities were to secure a warrant to enter a suspect's home to plant a listening device, it wouldn't make much sense to notify the suspect that authorities were entering the home to plant it.

It may be that in both cases where outrage exists over the angry librarians and sneak and peak provisions in the *USA PATRIOT Act*, very few members of the media and the public at large know much about the authorities that have existed for years in the law enforcement community. It wasn't until 9/11 and the scrutiny of the *USA PATRIOT Act* that awareness of authorities, at least the newly granted powers for national security investigations, began to grow. However, understanding the provisions in the *USA PATRIOT Act* is only part of the picture; they must be considered within the context of the common law tradition in American law enforcement. Once this is done, a view of the *USA PATRIOT Act* as more reasonable and more respectful of constitutional protections comes into focus. In fact, one wonders why it took a 9/11 tragedy for the U.S. to consider using existing law enforcement tools against terrorists, a much more menacing threat to society than the common criminal.

Checks and Balances

Although there is potential for abuse with legislation like the *USA PATRIOT Act*, many checks and balances exist to make this abuse less likely. For example, several of the new surveillance powers granted to law enforcement will "sunset" in four years, which means they will lapse unless reenacted by Congress. The increased list of offenses for the granting of wiretaps in Sections 201 and 202 expire in 2005. Additional sections that will sunset include 203(b), 204, 206, 207, 209, 212, 214, 215, 218, 220, and 223.

There is also legal recourse in the courts available to those opposed to the *USA PATRIOT Act* or any other of the Bush administration's efforts in the antiterror campaign. For example, take the case of deportation hearings. Shortly after 9/11, John Ashcroft gave orders to the chief immigration judge to close deportation hearings to the press and to the public in "special interest" cases, fearing that if al Qaeda gained access to the list of detainees, the group would gain valuable information on the U.S. antiterror campaign. Almost immediately this policy was contested in court, where civil libertarian groups and members of the media have had a chance to present their views.

The point is, whether you feel the antiterror campaign and efforts like the *USA PATRIOT Act* were overreactions to 9/11 or were very necessary to ensure security, the passage of the bill and its concomitant oversight show that the process of democracy is alive and well in America. Members elected by the people passed the *USA PATRIOT Act* with overwhelming majorities in both houses of Congress. Once the bill was signed into law, civil libertarian groups began using FOIA requests to scrutinize actions taken by the administration. When there are concerns, as there were with the release of the names of terrorist suspect detainees, the detaining of suspects at Guantanamo Bay, Cuba, or the use of the enemy combatant designation, opponents can take judicial action.

In the standard process of democracy, each of these cases eventually winds its way through the court system with an outcome that may or may not be favorable to the government's side. At the same time, Congress maintains its oversight role by holding hearings and gathering evidence on how the executive branch carries out the law.

The executive branch can even act as a check on itself. In 2003 the inspector general for the Justice Department issued a report critical of the administration's handling of illegal immigrants caught up in the post 9/11 sweep, claiming that some detainees with no connection to terrorism were held in captivity longer than necessary. William F. Schulz, executive director of the watchdog group Amnesty International USA, said in a *New York Times* article that the inspector general's office "should be applauded for releasing a report that isn't just a whitewash of the government's actions."[17]

These actions show the system of checks and balances working exactly the way the Founding Fathers intended. According to Jeffrey Rosen,

> In a series of court cases, federal judges have insisted on the importance of judicial oversight of the president's powers of detention and deportation. And in the debates over the *U.S.A. Patriot Act* and the homeland security bill, libertarians on the right have joined with civil libertarians on the left in persuading Congress to repudiate the Bush administration's more draconian proposals for expanded surveillance authority. In this sense, the greatest protector of American liberty during the past year turned out to be something so basic that we often take it for granted: the checks and balances provided by the separation of powers in the Constitution.[18]

Rosen suggests that the increase in powers to fight terrorism was far less sweeping than those passed in Europe. For example, in France police were given the power to search private property without a warrant. Paul Rosenzweig, a senior legal research fellow with the Heritage Foundation voiced a similar assessment: "The only significant diminution in personal freedom for American citizens is that we now have to take off our shoes at airport security."[19]

Sometimes it appears that many individuals have lost faith in the system of checks and balances that has held the world's longest running democracy together through scandals, impeachments, wars, and assassinations. As a reminder of how this system works, one should remember the Watergate scandal. The Watergate break-ins, secret funds, and dirty tricks, along with the subsequent cover-up, were a significant abuse of power by a sitting president. In another country without the constitutional safeguards present in the United States, a leader may have gotten away with these outrageous acts. However, even though it took some time, the system of checks and balances in the United States eventually caught up to Nixon, with the judicial and the legislative branches playing their designated roles. The Supreme Court checked the executive branch by requiring Nixon to release the tapes of Oval Office conversations. The legislative branch carried out inquiries through the investigative committees, which were vital to obtaining Nixon's resignation. Both Democratic and Republican members of Congress supported the recommendation by the judiciary committee to impeach the president.

After the Watergate scandal, various measures were implemented to prevent similar abuses from occurring in the future, including new codes of ethics adopted by the Senate and House of Representatives in

1977; the *Ethics in Government Act* enacted a year later in 1978; a special prosecutor law; and campaign contribution limits.

Government in an Open Society

It has become a matter of fact that nothing in the United States, especially an abuse of power, remains under the cover of darkness for very long. As David Brin has noted, only in America could an ordinary citizen like Paula Jones of Arkansas file a lawsuit against Bill Clinton that would trigger a cascade of events leading to the impeachment of the president. Or consider former Republican majority leader Trent Lott's suggestion at a birthday party for retiring senator Strom Thurmond that the country would be in better shape if it had elected Thurmond when he ran for president many years ago. Once the media began swarming over the story and Lott's history of supporting segregation leaked out, one of the most powerful men in U.S. politics was forced to resign his position as majority leader.

The Paula Jones and Trent Lott cases illustrate that the check and balance often forgotten about in our society is the openness that forces information out into the light of day. As the Watergate scandal showed, it only takes a couple of determined reporters on the scent of a story to help scuttle a presidency. In an example from the private sector, the book *24 Days: How Two Wall Street Journal Reporters Uncovered the Lies that Destroyed Faith in Corporate America* documents how the curiosity of two financial reporters covering Enron and their discovery of some hidden balance sheet irregularities set off a chain of events that caused the collapse of a mighty corporate juggernaut and resulted in the largest bankruptcy in U.S. history.[20]

It was Supreme Court Justice Brandeis's dictum that "sunlight is the best disinfectant," and in a society with C-SPAN recording every congressional speech and reporters from twenty-four-hour cable-news programs blanketing the nation's capital, a tremendous amount of sunlight is being directed at government by what Carlyle called the Fourth Estate. The Internet has further amplified this illumination with thousands of individuals reporting their observations and experiences on discussion boards and in the Web page diaries called blogs. Rarely does an item of interest slip past the ubiquitous attention of the many individuals interested in keeping the government in check.

Admittedly, the government has shown a historical resistance to greater openness; we see this with the Bush administration, which has been intransigent in resisting disclosure. Even so, it is becoming harder for administrations to keep their secrets locked behind a wall of privilege. When the prying eyes of the media combine with those of thousands of public interest groups and activists fighting for their various causes and of individuals posting news on the Internet at a moment's notice, you have a World Wide Web of watchfulness that even the government has difficulty resisting. Brin compares the many reporters, activists, insiders, and average individuals who search for abuses of power to cells of the immune system looking for a deadly disease: "In social terms, our contemporary neo-Western civilization already throngs with the human equivalent of T-cells, independent-minded persons who are well educated, skeptical, and driven by pumped-up egos to the point where their most devout goal is to find and reveal some terrible mistake or nefarious scheme."[21]

Unfortunately, this system appears to have an autoimmune deficiency where the T-cells sometimes get out of control and begin going after healthy parts of our body politic. In some cases, such as in the privacy/security debate, it seems that watchdogs find it easier to assail and demagogue the government than to engage in a thoughtful discussion about how to implement real solutions. Setting up Orwellian straw men may capture the public's attention, but it rarely helps find answers to difficult questions.

Frequently, attacks put the government in a no-win situation. After Timothy McVeigh and the Oklahoma City bombing, the government was condemned for not doing enough to infiltrate right-wing militia groups. After 9/11 the FBI was criticized for not connecting the dots on Arab visitors training in U.S. flight schools. After the D.C. sniper case, the government was ridiculed for being distracted by a white van that wasn't used in the shootings. However, when the government proposes a program like TIA to help piece clues together and to prevent future acts of violence, it is castigated for threatening civil liberties. In fact, the TIA project was not to be developed in the darkness of Big Brother secrecy, but in the daylight of openness, as was witnessed at a DARPA symposium in California, the notes of which were later posted on their Web site. DARPA also submitted multiple reports to Congress on TIA that outlined the active steps the agency planned to take to protect privacy and prevent abuse.

In arguing that the U.S. government is no Big Brother, this book in no way suggests that we should blindly trust our leaders with increased powers in the war against terrorism. Instead, it serves as a call to bring the debate back to a more rational level that avoids the rhetoric and emotionalism plaguing both sides. Setting up bogeymen in an attempt to shut down discussion does little to contribute to the democratic process and the goal of finding ways to battle terrorism while protecting important freedoms.

Some public-opinion data shows that over time the country has begun to reject the characterization of the U.S. government as Big Brother. An NPR/Kaiser/Kennedy School poll in 2000 showed that confidence in the U.S. government had reached its highest approval rating in a decade, with 51 percent of Americans confident that the government will actually solve the problems it encounters.[22] Many other polls taken since 9/11 have shown strong support for the government in the antiterror campaign.

In his 1946 essay "Politics and the English Language," Orwell was critical of those who so overused and misused words that they lost their meaning. How ironic it is that this has become the case with his best-known phrase.

CHAPTER 11

INVASION OF THE DATA SNATCHERS

Just when one thought a world dealing with the threat of Big Brother couldn't get any more frightening, enter the latest villain in the war against privacy, corporate America. Many individuals concerned about privacy suggest that the threat of corporations collecting data on Americans has become more insidious than that of the government. Garfinkel hints at a shift in focus by privacy advocates from Big Brother to corporate America:

> The future we are rushing towards isn't one where our every move is watched and recorded by some all-knowing "Big Brother." It is instead a future of a hundred kid brothers that constantly watch and interrupt our daily lives. George Orwell thought that the Communist system represented the ultimate threat to individual liberty. Over the next 50 years, we will see new kinds of threats to privacy that don't find their roots in totalitarianism, but in capitalism, the free market, advanced technology, and the unbridled exchange of electronic information.[1]

Data Collection by Corporations

There is no arguing that data is the lifeblood of the modern corporation. For example, it's likely that many of you reading this went to the grocery store this week and used a supermarket discount card when checking out. The card saves you a few dollars and allows the grocery store to learn a little about you. According to Catalina Marketing, which specializes in supermarket loyalty card programs, one third of the nation's thirty thousand supermarkets use them and many are mandatory.[2] Through the card, the store can create a history of all your purchases, which it can use for marketing purposes. For instance, if the store knows you buy health products, it may offer you a discount on a new line of vitamins.

Like the grocery store, virtually every company or organization you interact with creates an electronic record of your transactions. Richard Ericson and Kevin Haggerty highlight the wide range of data that is stored on individuals in a modern society.[3] These include data on personal credentials (birth certificates, driver's licenses); financial activity (ATM cards, credit cards, tax returns); insurance (health, home, and vehicular policies); social services (files relating to social benefits, health care, and pensions); utility services (data from telephone, cable television); real estate (purchase, sale, and lease agreements); entertainment (travel documents, theater tickets, Nielsen Ratings); consumer activity (purchase records, credit accounts, survey of consumer preferences); employment (applications, examinations, performance assessments); education (applications, records, references); and legal services (court records, legal aid files).

Although individual databases put many at unease, the more pressing fear is when companies combine distinct databases to create comprehensive profiles on consumers. Today many billion-dollar marketing firms and information brokers specialize in building large data warehouses of consumer data. In fact, any company with enough financial resources can create such a database because the sources of data are numerous. Information, including marriage, property, and mortgage information, can be obtained from list brokers, telephone companies, bank and financial databases, credit reports, and public records. In other cases, companies have joined together in the data-collection effort by forming holding companies, where affiliate members share information with one another under a single corporate umbrella.

Tracking Online Behavior

Critics also worry about the rapidly accumulating data on peoples' online activities. In a digital medium like the virtual world of cyberspace, it is very easy to capture and store a record, or "electronic footprint," of someone's activities. Unless you have been living under a rock and have never seen a modem, you are probably aware that one way this occurs is through the use of cookies, text files written to your computer with unique numbers that allow a Web site to know when you return.

Cookies are often paired with another technology called Web bugs. These are little pieces of code, the size of a period and hidden in the Hypertext Markup Language that powers a Web page. The Web bug is like a beacon that announces your arrival by reading your cookie, tracking your activities on the Web page, and then sending the information back to the server. The information may not identify you personally, but if you register with a Web site, the number in the cookie can be linked with your name, and each time you return, your identity can be determined.

Most people might not care if a Web site like CNN.com uses a cookie to present customized news headlines for regularly returning surfers. Some voice concern, however, when large Internet advertising brokers like Doubleclick compile cookies across Web sites and use them to create a profile of your surfing habits. For example, if you visit ESPN.com and Doubleclick provides ads for the site, the company will send you a cookie with a unique identifier. When you visit another Doubleclick ad-served site, perhaps Golf.com, a Web bug will send information from your cookie back to Doubleclick, indicating that you are the same person who visited ESPN.com. This allows Doubleclick to build a profile indicating that you are a sports fan that likes golf.

We see a similar phenomena developing in the offline world of digital video recorders (DVRs). Brought to market in 1999 by TiVo, DVRs allow viewers to record television on a hard drive, pause a live program, and skip commercials. Because the devices must connect to TiVo's servers to collect programming information, the company is able to capture information on a subscriber's viewing habits. It can then sell information, such as which commercials viewers are skipping, to advertisers, who can use it to make better advertising decisions.

Contrary to what its critics say, this kind of activity is not a result of the invidious intentions of TiVo or Doubleclick to spy on your behavior; rather, it allows advertisers to place spots that are more likely to interest you. Doubleclick wouldn't last very long as an advertising broker if it sent an ad on knitting instead of golf to the sports fan mentioned above. The individual also benefits by seeing an ad that is more interesting and relevant to his or her tastes.

At the moment Doubleclick's privacy policy claims that there is no personally identifiable information associated with your profile. However, some worry that in the future Doubleclick might pair online registration information with surfing behavior and thus threaten online anonymity.

An even greater concern is that once you are personally identifiable, your online profile could be tied to information about your offline habits. This almost happened in 1999 after Doubleclick completed a $1.7 billion acquisition of Abacus Direct, the country's largest catalog marketing firm. Doubleclick's plan was to link its online profiles with Abacus Direct's offline profiles. After pressure from privacy groups, the company backed down and promised to keep the identity of surfers anonymous, at least for the near future.

Online Conveniences

Even though much invective has been directed at companies that collect personal data, many consumers seem to turn a deaf ear to these worries. For every person who avoids the Internet due to anxieties about Doubleclick, many more seem willing to provide personal data online. Research from a privacy and American business survey showed that 68 percent of respondents were willing to provide personal data online to receive targeted advertisements.[4] Along with the online behavior of consumers, this suggests that many people think the conveniences they gain from technologies like online shopping far outweigh the risks to privacy. Anyone who ever has stayed up late shopping online at Christmas in his or her pajamas can testify to the allure offered by the Internet.

In another example of online convenience, single sign-on (SSO) services like Microsoft Passport are frequently criticized by the champions of privacy, but praised by consumers who like the idea of logging in once and having instant access to all the Web sites they visit during an online session. Passport stores personal information about a user, including credit card numbers and addresses and provides this information to Web sites of the user's choosing. Microsoft has reported that hundreds of millions of individuals are currently using the service.[5]

One of the appeals of online shopping is the number of products and services that can be made to order for the consumer who is willing to turn over a little personal data. Just as in the past, when the clerk at the local grocery store knew your tastes and might tell you to come back on Friday when your favorite fruit or vegetable arrived, so too can the Internet give you a personalized experience. Many Internet users are aware of the collaborative filtering techniques used by companies like Amazon.com and Wine.com. Their software builds individual profiles of visitors to a site and combines this information to create affinity groups of like-minded individuals that serve as the basis for making recommendations. At Amazon, the idea is that if you like a book, you may also like a different book that was bought by people with similar tastes to your own.

Personalizing the online experience is not limited to recommending books. For example, TiVo can autorecord television programs based on an individual's viewing habits. In the world of cosmetics, a Web site owned by Procter & Gamble named Reflect.com offers a range of customized beauty products for women. Visitors to the site complete a detailed questionnaire, which is used by Reflect.com to develop a complete beauty solution. Products can be personalized down to the fragrance, the packaging, even the name, based on responses from the questionnaire. This form of one-to-one marketing provides benefits to both parties in the transaction. The customer enjoys a more satisfying experience that is tailored to her personal tastes and interests, while the company more efficiently markets the right products to the right consumer.

The Free Flow of Information

Economists have long known that restricting information in the economy leads to greater costs for consumers. The economist George Akerlof won a Nobel Prize for his work of many years ago on the topic of incomplete information, what economists refer to as asymmetric information. In his paper "The Market for Lemons," Akerlof shows how a buyer's lack of knowledge about whether a used car is a lemon drives down the price of quality used cars.[6] Sellers therefore leave the market because they cannot get a fair deal, leaving relatively more lemons for sale.

The free flow of information can have significant benefits for many areas of the economy. A widely cited figure is from economist Walter Kitchenman, who claims that mortgage rates are as much as two points lower due to the free flow of financial information between companies.[7] For instance, large information brokers like Acxiom combine real estate and mortgage data gathered from deed filings at county court houses. Having home pricing and mortgage rate information in one place helps to create a very efficient market.[8]

Many have argued that greater privacy regulation, which would place restrictions on how companies use information, would have the effect of increasing costs in some markets. According to a study by the Information Services Executive Council cited by Sonia Arrison, an increase in privacy regulation would amount to a $1 billion information tax on consumers, as retailers would pass on higher costs of anywhere from 3.5 to 11 percent.[9]

To Opt or Not to Opt

Many readers may be thinking that a defense of the data collection practices of corporations leaves the consumer naked and without protection against the overwhelming might of the corporate marketing machine. On the contrary, just because a company knows something about a consumer, it is a non sequitur to think that company should have the right to harass the consumer with the equivalent of a marketing blitzkrieg. Counterintuitive though it might seem, I believe the most reasonable approach for the twenty-first century is not to direct our energies at restricting the flow of personal information between corporations, but to focus on how they use it, particularly in terms of direct marketing.

In an age when it will be next to impossible to hide one's interests, shopping habits, and contact information (mailing addresses, cell phone and pager numbers, and e-mail addresses) from profit-driven corporations and businesses around the world and in which the cost of communications continues to drop, it will become necessary to adopt an opt-in approach for direct marketing. Opt-in is a term that comes from the world of e-mail marketing where companies only send messages if the consumer explicitly signs up for them. This is the opposite of opt-out, where companies send out marketing messages until an individual proactively requests that his or her name be removed from the mailing list. As we've seen with spam, opt-out tends to be the norm on the Internet, where individuals are bombarded with a never-ending stream of unrequested marketing messages from companies.

An opt-in approach to direct marketing will be one way to carve out some personal space for an individual in the open society of the twenty-first century. It will send a message to spammers that unrequested e-mails cluttering inboxes on personal computers will not be tolerated. It will also make it clear to telemarketers that the sanctity of someone's home is off limits and that a family's dinner should not be disturbed by a telephone call from an eager salesperson. To be consistent, opt-in might even be extended to junk mail, allowing people to choose if they want to receive the latest catalog from J. Crew or not.

In fact, popular support for a national do-not-call database shows that many Americans prefer to have control over whether they receive direct marketing messages. Can calls for a national do-not-e-mail and do-not-snailmail database be very far behind? Of course, before such a database would be effective in reducing the amount of spam, for example, improved methods for identifying the culprits will have to be in place.

This is not to say that all forms of marketing should be strictly limited. Marketing is an essential part of the free enterprise system, allowing companies to reach out to consumers. However, there are many different advertising venues, and in an opt-in world, marketing professionals will have to wean themselves of their dependence on direct marketing and explore new channels to reach consumers. The use of DVRs with their ability to skip television commercials has already sent the message that marketers need to adapt if they are to survive. One result has been an increase in product placements in movies, television shows, and other forums. Opt-in may even force marketers to improve their messages so that they can convince people to opt in, in the first place.

One may wonder whether it will be possible to staunch the tide of marketing messages in the expanding information revolution. We have to keep in mind that the explosion in information technologies cuts both ways. Whereas corporations are better able to collect details on our personal lives, improved technology allows consumers to better communicate their preferences to corporations and to the regulators who would enforce the opt-in rules. There is simply no excuse for a corporation to call someone mistakenly when database technologies make it easy to cross-check names and numbers in a do-not-call database. At the same time, greater transparency will enable the identification of those who flout the rules.

Data Exchange as Gossip

Some scholars, such as Eugene Volokh of Stanford University, have made the case that when companies exchange data with each other, they also are exercising their right to free speech.[10] In this way, the use of data by corporations is a lot like gossip.

Solveig Singleton suggests that most of the consumer information held by corporations is not much different from the kind people exchange with one another.[11] If you tell a coworker that you are buying a new car, the information may make it around the office much as the details of the car purchase might be passed from the automobile dealer to another company. In both cases, the individual has very little control over the data in a third party's hands. Singleton makes this point with an example of someone buying a lawnmower from Sears: "Why should the information about the sale belong only to the customer and not to Sears as well? If the customer were to complain about the transaction to Consumer Reports, he would not have to ask Sears' permission. Why cannot Sears boast of the transaction to its creditors?"[12]

Some have argued that databases should not be equated with gossip because they are more harmful. Jeffrey Rosen claims that unlike gossip, information captured online and in databases can last forever, and in the anonymous world of the Internet, harmful information can be difficult to counter.[13] However, as anyone who has gone to a small school or lived in a small town knows, gossip can be just as deleterious. Singleton says, "But one cannot meaningfully distinguish consumer databases from gossip on the grounds that gossip causes no harm. Historically, gossip exchanged within small communities could cause terrible harm indeed, because public commentary within those communities had powerful influence over others' lives."[14]

She goes on to quote an anthropologist who says, "When individuals are dependent on one another for cooperative hunting, farming, herding, or for access to wage labor, gossip and the reputations it creates can have serious economic consequences."[15]

Kevin Kelly of *Wired* magazine admits that there was little expectation of privacy in villages or towns. The difference, he says, was that you knew just as much about other people as they knew about you. Today, he claims there is an imbalance of information. "There was a symmetry to the knowledge. What's gone out of whack is we don't know who knows about us anymore. Privacy has become asymmetrical."[16]

Although we might not know the name of every company that has information about us, a mitigating factor is that unlike in the case of village gossip, consumers are relatively anonymous to corporations. When John Doe is in a database of ten million consumers, most employees for the company don't have a clue as to who John Doe is. However, in the village, John Doe is no longer a John Doe. Everyone knows John.

It is also likely that corporate data is more accurate than gossip could ever be. For one, companies have an incentive to make sure that their data are as correct as possible so that they can use them effectively to target consumers. People who spread gossip have no such incentive. In fact, the more they exaggerate a story, the more appealing it is likely to become.

Consumer data resembles gossip in another way: People frequently have very little expectation of controlling it. Virginia Postrel has made this point:

> The other party in any relationship—whether your former landlord, your boss, your ex-girlfriend, or Amazon.com—owns information about you as surely as you do. Gathering and sharing such information is as old as gossip and is absolutely essential to a free society. Neither speech nor commerce can function if such communication is illegal. Privacy advocates want to outlaw not only journalism but reputation.[17]

Courts have long supported the notion that people do not control the information generated by many of their transactions. For example, in 1974 the Supreme Court decided that your bank records belong to the bank, which can use them for whatever purposes it chooses. Of course, Congress can decide to step in and place limits on the exchange of information. The *Gramm-Leach-Bliley Act* requires financial institutions to disclose to customers their policies for protecting the privacy of personal information and to allow them to opt out if they choose.

Perhaps one day individuals will have more control over their personal data if systems of pseudonymity and anonymous credentials become technologically feasible and widely available. Just like gossip, corporate marketing would slow if its purveyors found it impossible to get the "scoop" on consumers. The fact that corporations need consumer information to better tailor products and advertising pitches to their customers suggests that they would be highly resistant to such a system. Furthermore, it is questionable whether the public would demand anonymous credentialing services, especially considering that many software companies offering privacy-enhancing software are struggling to stay afloat.

Personal Data, Private Property, and Copyrights

A popular belief is that when a person interacts with a company or organization, data from the transaction should not move outside the relationship and be used for another purpose without the individual's consent. This idea stems from the principles in the *Code of Fair Information Practice* discussed earlier. Taking this a step further, many argue that personal data is the private property of the individual and should be protected as such. This was one theme of Anne Wells Branscomb's 1994 book *Who Owns Information? From Privacy to Public Access*.[18] The logical conclusion to this argument is that if marketing companies are going to profit off of someone's personal data, that person should be compensated. In some ways, this would result in giving personal data the same kinds of protections as apply to copyrighted information.

Copyrights have long enabled creators such as authors and musicians to be compensated when their work is used. The use of copyrights has a long history in this country, going back to the Constitution, which says, "the Congress shall have power . . . to promote the progress of science and useful arts, by securing for limited times to authors and inventors the exclusive right to their respective writings and discoveries."

Provisions provided for by the Founding Fathers were influenced by English laws, such as the *Statute of Anne*, which Parliament enacted in 1710 to address the concerns of English booksellers and printers. Copyrights protected the English booksellers like they protect creative types today by ensuring that they can be compensated for their works.

Copyrighted data, just like personal data, is challenged by technology that facilitates its distribution. The rampant spread of file trading on the Internet is the clearest example of a technological threat to copyrights. Kazaa, the P2P file-trading software and descendent of Napster, connects a user to millions of other computers around the world, making it possible to download virtually any song, movie, or piece of software without ever having to pay a cent to the creative talent who brought it to life.

The *Digital Millennium Copyright Act* (DMCA) was passed in 1998 in an attempt to restrict technologies such as P2P that allow people to circumvent copyright laws. Civil libertarian groups have opposed the act in court, arguing that it puts restrictions on the fair-use doctrine, which allows consumers to make copies of copyrighted content for personal and other "free-speech" types of uses, such as for criticism, teaching, or research.

The fair-use doctrine wasn't much of a concern to the publishing industry until advances in technology began to undermine it. Although it was possible, bootlegging eight-track tapes and cassettes could not be accomplished very efficiently or cost-effectively on a mass scale. In most cases, fair-use worked as intended, enabling people to make copies for themselves, while not infringing on copyrights.

Digitized data and the connectivity of the Internet have changed all of this. With digital recordings of music, individuals can now create an MP3 of any commercial song and make it available for millions of other users to download in just a few seconds with a cable modem.

Defending technologies like Kazaa within the fair-use doctrine represents a serious stretch of the imagination when the sole purpose of the software is to allow millions of people to get around copyright restrictions. There are plenty of ways to engage in fair-use without using software designed to distribute copies of creative works to Internet users around the globe.

The battle by civil libertarian groups against the *DMCA* demonstrates inconsistencies in their position. First, they fight tenaciously to protect personal data at any cost, but reject the idea of safeguarding copyrighted information against online marauders. In their hue and cry lamenting the impact of technology on the spread of personal data, nary a peep is heard about its damaging consequences for copyrights and their owners.

Their answer to this apparent contradiction is that online users who circumvent copyright protections are exercising their constitutionally protected right of free speech. However, when companies want to use their free speech rights to market data, they rise in protest. This inconsistency shows up in other arguments by civil libertarians, such as in the position of supporting door-to-door canvassing, but not mailbox-to-mailbox marketing.

It may be that efforts to fully protect copyrights will become moot as the raging river of information grows ever more uncontrollable. There may be occasional successes in harnessing channels of this fast moving flow of data. One approach has been to offer market-based alternatives like Apple's iTunes network, whose downloadable songs are offered with such extreme convenience, high quality, and low prices that some file traders are converted.

In other cases, the industry has gone on the offensive by targeting individual file traders with lawsuits and flooding networks with mislabeled or degraded files. Large hardware and software companies like Microsoft, Intel, and Hewlett-Packard have also gotten into the act by building digital rights management into the very hardware of computers, enabling encrypted media content to be extracted only on "trusted" machines.

Yet, even with all these efforts, it seems ever less likely that the rush of copyrighted information across the Internet can be contained. At some point digital content has to be viewed by the human eye or listened to by the human ear; thus, it can always be captured by some recording device and distributed through the ever-expanding number of communication mediums. Control of these channels seems less likely as well, especially as P2P technologies are refined and anonymous models of file sharing are introduced in an ongoing arms race with the media industry.

What will creators do if they cannot receive compensation for their works? China, which ranks as one of the worst countries for protecting copyrights, offers a vision of how artists may survive in a world of irrepressible file trading. Instead of relying on revenue from music albums, music stars have been forced to look to other sources of revenue. Their music recordings can serve as a vehicle toward individual fame, which can then be parlayed into packed concert halls, radio station revenue, merchandising deals, advertising contracts, and sponsorships. This attitude is reflected by Chinese musician Wang Lee Hom, quoted in an article by the *International Herald Tribune*: "Until they pirate my body, I can rely on personal appearances. I am forced to view albums as only a promotional tool."[19]

Preventing the spread of voluntarily disclosed consumer information in the coming years seems even more unlikely than protecting copyrights. Courts have consistently ruled against the idea of individuals owning their data in the sense of owning property. As Arrison points out, "In the United States, there is no right, historical or otherwise, to stay in complete control of your information after giving it away. Any 'property' right is relinquished once an individual voluntarily puts personal information in the public domain."[20]

The more significant issue is the potential impact on society of restrictions on personal data. According to privacy expert Roger Clarke, "Broadly enforced, such a property right would be antithetical to an open society. It would pose a threat not only to commerce, but also to a free press and to much political activity, to say nothing of everyday conversation."[21]

How would the media react if they had to get permission every time they mentioned a fact about a person in a story? What if politicians were prevented from talking about their opponents due to personal data laws? Where would it stop? One might imagine that even speaking about a person in public could risk a personal-data lawsuit.

The irony is that in a world with more privacy than the village, where people can easily disappear into anonymity, businesses need data about their customers if they are to do business with them. Stephen Nock argues that although we have more privacy today, the cost of that privacy is that we have to be willing to turn over some data about ourselves.²² For example, how can a bank grant someone a credit card if they have no idea who that person is?

> Computerized systems that permit others to investigate our employment or financial background are gaining widespread popularity among employers or creditors. Although some may decry the establishment of computerized records of individuals as a "loss of privacy", it would be more correct to see such developments as the cost of vastly expanded amounts of privacy. Indeed, there would be little need for massive databases on individuals were there no privacy. . . . If we knew everything about everyone, there would be little reason to collect and store the details of their biographies. It is only because major portions of our everyday experiences are legitimately (often legally) defined as beyond scrutiny that distrust can arise. It is, in other words, only because we enjoy such great privacy that surveillance arises in the first place. To enjoy some degree of predictable social order, we may have either privacy and surveillance, or little privacy.²³

Rather than an invasion of the data snatchers, the collection of personal data by companies is a necessary component of the modern, information-age economy, one that most consumers grudgingly accept. Even the Cypherpunks, an infamous group of cybergeeks, coders, and Internet pioneers with a long history of fighting for electronic privacy, acknowledge this principle. According to Eric Hughes in "A Cypherpunk's Manifesto,"

> We cannot expect governments, corporations, or other large, faceless organizations to grant us privacy out of their beneficence. It is to their advantage to speak of us, and we should expect that they will speak. To try to prevent their speech is to fight against the realities of information. Information does not just want to be free, it longs to be free.²⁴

Tighter regulations on corporate data would please one unexpected beneficiary: For terrorists who want to hide their electronic transactions in a sea of anonymity, controls on personal data are a boon. Investigators will have a much harder time staying on their trail if it becomes more difficult to piece together data locked away in corporate silos. Only through access to data will IA programs, like the Treasury's FinCEN, which look for financial crimes, or other programs being developed to fight terrorism, be able to serve the public—something that a new privacy regulatory regime could jeopardize.

CHAPTER 12

INFORMATION DOES NOT KILL PEOPLE; PEOPLE KILL PEOPLE

An underlying assumption in the privacy debate is that if personal data could be secured properly many of the worst threats to privacy would be a distant memory. Whether a privacy concern involves a case of medical records being revealed, financial information being stolen, or online footprints being sold, personal data is caught in the crosshairs. As a result of this focus, those most zealous about protecting privacy apparently wish that personal data could be hidden away, locked in a Fort Knox–type fashion with the strongest available encryption technology.

David Brin spent much of *The Transparent Society* arguing that this kind of thinking was shortsighted and ultimately flawed. No matter how much security is used, the elites in society will always have the better technology and a greater capability to access protected information. Even if one encrypts data with the strongest standards available, there is always the risk that tomorrow's computers will expose its secrets to the world. Brin suggests that our best hope is to demand transparency from those who have the keys to our information to make sure they don't misuse it.

The Nosy Neighbor

Even if the Fort Knox fantasy somehow became a reality and scientists developed a way to secure data, perhaps using quantum encryption or some other exotic technology, there would still be one problem—human beings.

A misconception in the privacy debate is that the spread of personal data is the result of computers in the information age. There is no arguing that technology helps facilitate the flow of information through society. However, information isn't limited to just the kind running through fiber optic lines. We often forget in this age of unlimited storage and light-speed data transfer about that other great reservoir of personal data—the human brain.

In a society in which people interact with many others during any given day, human beings are just as likely to spread personal information as computers. This hearkens back to the analogy of data exchange as gossip. Just as Amazon.com can learn that you like romance novels, so can the lady at the local bookstore. I call this the Nosy Neighbor Principle. Even though people are more spread out across the country today than in the past, you typically encounter many of the same people on a regular basis. These people know things about you and usually aren't afraid to share these things with others.

The nosy neighbor doesn't have to be a neighbor. This person can be a policeman, the lady at the local grocery store, someone at the post office, a coworker, and so forth. Anyone who has seen gossip spread among these networks knows that there is some truth to the idea that you are only a handful of relationships away from anyone else in the country. As a result of the Nosy Neighbor Principle, privacy efforts that try to stop the spread of data are usually doomed to fail.

No matter how hard one tries to secure electronic data, personal information is still going to be vulnerable whenever humans are involved. In many cases, before data is stored in a secure computer, a person must first enter it. As a result, while information is being locked away in a database table, it's also being stored in the human equivalent of a database, the human mind.

Even if the nosy neighbor didn't enter your personal data into a database, he or she may have seen you engage in a transaction. For instance, my wife and I just finished buying a house in Virginia. As anyone who has gone through a home purchase knows, it is a complex transaction that involves a number of parties, including the buyer, seller, agents for both parties, the mortgage broker, settlement lawyer, and so forth. Now that we have moved into our new home, our real estate agent lives a few blocks away and the seller's agent just a few doors down. Although anyone interested in the price we paid for the house could visit the local courthouse and look up the record, I think there is a good chance that gossip in my neighborhood will make the trip unnecessary.

Many people will argue that the nosy neighbor pales in comparison to data being zapped through global networks and stored in the gargantuan digital warehouses of large marketing firms. It is certainly true that information technology enhances the likelihood that multiple parties will exchange our personal data; however, you and I are just faceless individuals in a database of millions, which assures us a certain amount of anonymity. In the small, informal networks of people we encounter in our daily lives, however, personal data has the most significant impact on privacy. Who are we most concerned will find out that we have a medical condition or financial problem: a corporation based in Asia or our neighbor next door?

As I write this, two examples in the press illustrate the nosy neighbor problem. In the first case, we see the difficulty the government has in keeping information confidential. Bob Woodward has just written the book *Bush at War*, which chronicles the first one hundred days of the Bush administration's response to the attacks of 9/11.[1] The book is based on notes from over fifty National Security Council and other high-level meetings and interviews with over one hundred people. The president even agreed to do a couple of interviews with Woodward to make sure his perspective was covered. Insider information, from the highest levels of government, resisted secure databases and confidentiality agreements and made its way into Woodward's book.

In another example from the private sector, a few big retailers, including Best Buy, Target, and Wal-Mart have threatened legal action against several Web sites that post upcoming sales from the retailers. The retailers are irate that discount shopping sites such as FatWallet.com are able to get inside information about what products will be on sale for what price before the retailers announce it. Unable to control leaks of information, the retailers have threatened these sites with lawsuits charging copyright infringement. These examples and countless others show that both the private and public sector have trouble keeping information locked down and away from nosy neighbors.

Privacy advocate Simson Garfinkel gives a personal example of an experience with the Nosy Neighbor Principle.

> When I started dating my wife in 1993, we went together to get tested for AIDS at Boston City Hospital. The clinic was one of several in the city specifically set up to allow for anonymous testing. The nurse who took my blood had no idea who I was and never asked for any identification. She gave me a control number when I left so I could learn the results. But when my wife and I returned a week later, a woman who was volun-

teering at the clinic recognized me from a class we had taken together at MIT. Should that volunteer have been legally prohibited from telling people that she had seen me at the clinic? What about other people who happened to be in the waiting room who might have recognized me?[2]

This is not to say that we shouldn't attempt to restrict information in databases or that access mechanisms, audit trails, encryption, and other forms of security should not be put in place to make it difficult for people who would pilfer data in order to cause mischief. The point is simply to recognize that when you combine technological advances like high-speed networks and XML with old-fashioned human methods of data transfer like gossip, keeping a lid on information is next to impossible. It's like the boy who tried to plug the dike. There are just too many holes to plug, and data are leaking out everywhere. If we want to be more effective in protecting people from harm in the twenty-first century, our efforts would be better spent not by trying to lock down data, but by prosecuting the actions of people who use data to cause harm to others. The following examples highlight this point.

Don't Blame the Data

In many of the arguments supporting greater regulation of data, it's common to hear some anecdotal story about the abuse of personal information sure to instill sympathy in the reader. One frequently cited example revolves around medical data, such as cases of people who were fired after their employer obtained medical data and discovered the employee had a preexisting medical condition. The implication in such cases is that although the employer is clearly in the wrong, the availability of information is the root of the problem. The solution suggested by this reasoning is that if only we could find a better lock for information, incidents like these would be prevented.

There is no doubting that in medical discrimination cases, access to information can facilitate the regrettable and egregious act perpetrated by the employer. However, the fault does not lie with the employee's heath data located in some company file. By itself, that information can serve a number of useful purposes, such as guiding medical decisions in the case of an emergency at the workplace. The issue is when someone actively uses the information to cause harm to someone else, in this case, to discriminate (e.g., to deny employment to someone based on a medical condition). Fortunately, the remedy for this type of situation exists in laws that protect people from this kind of discrimination. For example, the 1990 *Americans with Disabilities Act* provides legal protection against health-related discrimination and the 1996 *Health Insurance Portability and Accountability Act* (HIPAA) protects consumers from losing health insurance if they change jobs, and it forbids discrimination based on "health status."

With recent advances in genetic testing, a similar workplace concern has been genetic discrimination. Take the case of Terri Seargent. She filed one of the first lawsuits for genetic discrimination, claiming that a North Carolina insurance company fired her after they discovered she was genetically predisposed to a lung and liver disease.[3]

Here again, Congress has recognized a concern with discriminatory action being taken against someone because of medical data. A bill entitled the *Genetic Information Nondiscrimination Act* would prevent health insurers from using genetic information to deny, cancel, or change the rates for health insurance, as well as prohibit the use of such information for employment-related decisions such as hiring or firing.

States have also taken measures to protect employees from genetic discrimination. According to the National Conference of State Legislatures, Wisconsin was the first state to ban genetic testing and discrimination in the workplace in 1991, and as of July 2002, genetic nondiscrimination in employment laws are in place in thirty-one states.[4]

Although it is true that legislation such as *HIPAA* does include provisions to protect the privacy of data, the Nosy Neighbor Principle shows that it can be very difficult to prevent people from uncovering personal information. This is especially true of medical conditions in the workplace, where people often show symptoms of an illness. Other times, an employee may share information with a coworker and find that the details end up spreading around the office. In modern work environments where people are crammed into open cubicles, it can be very easy to overhear a conversation on the phone or between coworkers that might reveal intimate medical facts.

Washington Post journalists Martha McNeil Hamilton and Warren Brown describe a good example of this in the book *Black & White & Red All Over: The Story of a Friendship*.[5] In the *Washington Post* newsroom, Hamilton sat next to Brown and overheard that he had a serious kidney ailment. Eventually, the two became friends and Hamilton ended up donating her kidney to Brown. Although the book describes an inspiring crossing of the racial divide (Hamilton is white and Brown is black), it is also a good example of how little in the workplace is private and how information that becomes public can sometimes have beneficial effects.

Health care is another environment where it is exceedingly difficult to keep information under wraps. In the well-known case in the 1980s involving Arthur Ashe, the hospital where he was receiving treatment leaked the fact that he had HIV. Of course, anyone going into a hospital, including Arthur Ashe, should expect his or her privacy to be respected. Amitai Etzioni suggests numerous technologies to protect medical information, such as layered security, auditing of access to records, smart cards to limit access, and fixed payments to insurance companies not linked to treatments types.[6] However, this doesn't prevent the nosy neighbor problem. There is always the chance that you will run into someone that just can't keep his or her mouth shut.

In the case of a public figure like Arthur Ashe, it's likely that he was recognized by a large majority of people in the hospital. In the 1980s HIV was relatively a new phenomena, and the fact that a public figure had been infected probably caused quite a stir in the hospital. A hospital may have the best privacy practices, the most secure databases, and the strongest confidentiality agreements, but if a nosy neighbor decides to gossip, the information will find its way out.

Despite the practical difficulties of locking down information in the health-care community, provisions in *HIPAA* that went into effect in April 2003 show that Congress is stubbornly determined to give it a try. Through an encyclopedia-sized collection of regulations that is expected to cost a whopping $17.6 billion over ten years[7] (although the HHS has made a dubious argument that greater use of electronic data will save money), health providers, insurance companies, and pharmacies are required to protect patient privacy or face steep fines and possible imprisonment.

For privacy advocates critical of the invasions of "Big Brother," the *HIPAA* regulatory regime drips with irony. In Orwellian fashion, it spells out in exhausting detail how hospitals and doctor's offices should run their businesses. In order to avoid the privacy police, office workers fret over how to hang charts on the walls, guard the fax machine from unwanted visitors, and speak in hushed tones on the phone. Some providers are even remodeling offices so that lingering patients won't be able to glance at unauthorized information. The rules are so misguided that if a patient mistakenly chooses a "do not announce" status, the hospital may not be able to let a family member know the person was admitted.

It's not just in the world of medicine that data is blamed for the wrongful actions of others. In cases of violence where information is used to track down and harm or kill a victim, it is easy to fall into the trap of faulting the data. One well-known example is that of Rebecca Schaeffer, an actress who was murdered by an obsessed fan. The fan hired a private investigator to track her down using state driver's license records, and in 1989, after discovering her address, he shot her to death at her home. This led Congress to pass the *Driver's Privacy Protection Act*, in 1994, which forbids states from releasing personal information from driver's license records.

Another case involves John Britton, a physician at an abortion clinic in Pensacola, Florida. Antiabortion protesters used his license plate to locate his address through state driving records and then posted it on the Internet in the form of a "wanted poster." He was ruthlessly shot to death in his home by an antiabortion protestor in 1994.

Many are quick to blame tragedies like these on the fact that the address of the victim did not remain private and thus left him or her exposed to a cold-blooded killer. Once again, going after the data is misguided. It is ludicrous to think that in today's modern information society, someone could ever fully hide his or her address. Think of how many business transactions require this piece of information. Unless they want to be like Theodore Kaczynski (a.k.a. the Unabomber) and live in a shack in the woods, most people can't just disappear. Even the Unabomber was eventually discovered.

If a person is determined enough to track someone down, he or she can find any number of ways to do so. As in the case of Rebecca Schaeffer, one could hire a private detective. Even with driver's license data off limits, the detective could use a number of other methods to get her address. For example, the detective could visit the movie shoot and follow her home. He could ask people around the movie set for information. With the Internet one can use any number of online services to search for people using court records, military records, voter registration records, professional license information, or even phone numbers, just to name a few sources.

> New "Phone Book" Raising Serious Privacy Issues Palo Alto, CA—Alarmed by the "ever-shrinking security and rights of individuals in the information age," the Palo Alto-based group Citizens For Privacy is calling for strict controls to be placed on "phone books"—printed directories of all the telephone numbers in a specified area. "With this new piece of technology," CFP head Nadine Geary said, "anyone could know your phone number in literally seconds." Exacerbating the situation, Geary said, is the fact that, in many cases, the subject's address is also printed right next to the number. "If this device is allowed to be distrib-

uted," Geary said, "literally anyone would be able to track you down at any time. It's frightening."[8]

There is still some uncertainty as to how the courts will eventually rule in situations where the uncovering of information precedes a violent act. In the case of *Amy Lynn Boyer v. Docusearch, Inc.*, the New Hampshire Supreme Court ruled that Docusearch, an Internet information broker, was liable for passing information to Liam Youens, a twenty-one-old obsessed with Boyer, who shot her as she left work one day.[9] The court found that Docusearch was liable because it had obtained the information under false pretenses, using "pretext calling" to trick Boyer into handing over her employer's address. One wonders how the court would have responded if the company had gathered the information in a more forthright and honest manner.

The court appeared to answer this question by claiming that Docusearch had a duty to Boyer to make sure the information it sold about her was used for legitimate purposes.

> The threats posed by stalking and identity theft lead us to conclude that the risk of criminal misconduct is sufficiently foreseeable so that an investigator has a duty to exercise reasonable care in disclosing a third person's personal information to a client. And we so hold. This is especially true when, as in this case, the investigator does not know the client or the client's purpose in seeking the information.[10]

Although an appeal is underway in a federal district court, the ruling by the New Hampshire justices could set a dangerous precedent. Requiring that every company that passes on information to another party be able to discern how that information will be used has the potential to stultify economic activity and hurt consumers. It would have a stifling effect on any transaction where data are exchanged, including companies checking credit histories, small businesses and nonprofit organizations reaching out to consumers, or health-care providers releasing information on a patient's health.

The most reasonable solution to cases like Amy Boyer's and the others mentioned above is not to outlaw the exchange of information or to pass stricter privacy regulations, but to enforce and strengthen already existing laws that protect citizens against harm. Going after a so-called invasion of privacy only distracts us from the real cause of the problem, which in this case is violence.

Just as "guns don't kill people, people kill people," information alone is not the problem. As Solveig points out, just because information can be abused doesn't mean you should outlaw information. Libraries contain information about nuclear weapons, but we don't ban libraries.[11]

Let me use an analogy provided by one privacy advocate. Lauren Weinstein has argued for the right to use a cell phone while driving a car.[12] He suggests that instead of creating new laws that punish people for using a cell phone while driving, we should enforce existing driving laws. If you cause an accident, his reasoning goes, you should be held accountable by laws currently on the books. Perhaps Weinstein and others should think about applying this logic to the privacy debate and stop calling for new privacy laws when adequate laws to protect the public from harm already exist.

Is Privacy Really the Problem?

In fact, this reasoning suggests that many supposed violations of privacy can often be reconsidered as violations of other, more primary rights. This returns us to the idea of the expanding definition of privacy. As the notion of privacy becomes bloated, it categorizes many offenses as privacy issues instead of allowing them to stand and be prosecuted on their own merits. So for instance, instead of classifying ogling as what it has always been at its core, a case of sexual harassment, it is remade into a case of privacy invasion. Instead of considering identify theft a security problem, it is redefined as a privacy issue.

As we saw earlier, the notion that privacy is built upon more fundamental rights can be traced back to reproductive cases like *Griswold*, which constructed a concept of privacy out of reproductive rights. Scholar Judith Jarvis Thomson has used a similar line of reasoning to question the very nature of the right to privacy:

> The question arises, then, whether or not there are any rights in the right to privacy cluster which aren't also in some other right cluster. I suspect there aren't any, and that the right to privacy is everywhere overlapped by other rights. . . .[13] For if I am right, the right to privacy is "derivative" in this sense: it is possible to explain in the case of each right in the cluster how come we have it without ever once mentioning the right to privacy. Indeed, the wrongness of every violation of the right to privacy can be explained without once mentioning it. Someone tortures you to get personal information from you? He violates your right to not be tortured to get personal information from you, and you have that right because

you have the right to not be hurt or harmed—and it is because you have this right that what he does is wrong.... In any case I suggest it is a useful heuristic device in the case of any purported violation of the right to privacy to ask whether or not the act is a violation of any other right, and if not, whether the act *really* violates a right at all. We are in such deep dark in respect of rights that any simplification at all would be well worth having.[14]

Thomson goes too far in her analysis by peeling away the layers of privacy until there is nothing left. On the contrary, there are clearly times, such as in the respite of one's home or during private communications, when a person's expectation of what we call privacy is more than reasonable. In this zone, few people would dispute the fact that the Fourth Amendment is a bulwark of defense.

However, Thomson is on track when she suggests that the issue of privacy has become so all encompassing as to render the concept nearly meaningless. It is virtually impossible for anyone from the government, corporations, Web sites, hospitals, to a coworker, to even a neighbor not to violate someone's privacy when the notion is defined as nothing short of looking at a person.

The danger is that when issues like fighting terrorism get caught in the grasp of this many-headed privacy Hydra, we become distracted from what is really at stake—whether there has been an abridgement of a person's right to act. Most critical is not whether Attorney General John Ashcroft knows that I checked out a book on nuclear weapons from the local library, but whether he uses that information to restrict any of my freedoms. This isn't to say that someone's privacy at the library isn't important, but rather it pales in comparison to the question of whether someone is prevented from reading in the first place.

In the conclusion of *The Limits of Privacy,* Etzioni hits on this point when he discusses the ways in which the line between what he calls scrutiny (privacy) and control (private choice) is often blurred.[15] He traces this confusion back to the very creation of a modern right to privacy in the *Griswold* case where the court sent mixed signals on whether restricting the use of contraceptives was unconstitutional based on scrutiny (the government entering the bedroom) or control (the government barring their use). I would add that the two concepts have been further muddled by the idea that privacy is akin to liberty, as Chapter 8 discusses.

This chapter suggests that the conflation of privacy and private choice directs unnecessary attention at the former at the expense of the latter. We see this misguided ordering of priorities in many issues where privacy is a concern. For instance, in coverage about TIA in the media, the discussions usually centered on what information would be collected on public citizens (privacy), rather than how the program might restrict behavior (private choice). We would be much better off if our attention were not directed solely at whether the government knows I like to travel to Europe, for instance, but whether it prevents me from doing so.

In fact, as Etzioni suggests, neglecting the issue of private choice in favor of privacy has the unintended consequence of exposing citizens to greater control. For example, when encryption locks communications behind a wall of secrecy, the government is forced to use more invasive procedures during its investigations. Or when identities are hidden in a veil of anonymity, citizens are forced to endure greater security hassles.

Etzioni's answer to this dilemma provides a vision for how privacy might be redefined in the open society of the twenty-first century. He advises us to return to a conception of privacy based on the foundation of the Fourth Amendment, where although the focus is on privacy rather than private choice, it is a more narrowly defined view of privacy with clear guidelines on when it must be respected and when it can be limited for the common good.

Etzioni also suggests that the amendment's restrictions on government scrutiny might serve as a model for private-sector interactions. Although I would agree with the spirit of his view, I would argue that our focus is properly directed at the government because that is where we have the most freedom to lose. In many cases, an invasion of privacy in the private sector might engender discomfort, such as the husband whose affair is uncovered after his wife rummages through his travel records, but rarely does it restrict choice, and when it does, there is usually legal recourse. I would return to John Woodward's point that feelings do not translate into a legal right and add that we hesitate to look to Congress to legislate based on them.

Once the notion of privacy becomes more circumscribed, we may be less tempted to disregard the issue of private choice. So for instance, if a Fourth Amendment interpretation of privacy gives the D.C. police the legal authority to use surveillance in public in the war against terrorism, we can expend less energy hand-wringing over the issue of privacy and more on ensuring that officials don't use their power to restrict the rights of activists on the National Mall. Applied to the other tools of openness, such as national identification, facial recognition, and IA, this perspective allows us to set aside doom-and-gloom speculation over the issue of privacy to focus on the real threats to freedom

Some may argue that I've drawn too clear a distinction between the issues of privacy and private choice; they might suggest that although it is true that in many cases the exposure of private data doesn't necessarily imply constraints on personal action, there are instances when it discourages it. The courts often prohibit any release of personal data that might inhibit someone from participating in First Amendment activities, often referred to as the chilling effect. For example, in *Tattered Cover, Inc. v. City of Thornton* (2002), the Colorado Supreme Court ruled that individuals have a right to purchase books anonymously under the First Amendment.

> When a person buys a book at a bookstore, he engages in activity protected by the First Amendment because he is exercising his right to read and receive ideas and information. Any governmental action that interferes with the willingness of customers to purchase books, or booksellers to sell books, thus implicates First Amendment concerns. Anonymity is often essential to the successful and uninhibited exercise of First Amendment rights, precisely because of the chilling effects that can result from disclosure of identity.[16]

Although I would agree that courts like the one in Colorado have the best intentions in mind when they try to encourage free speech, I believe their efforts to intuit a person's feelings and motivations add to the confusion that exists between privacy and private choice. The Colorado Supreme Court indicates that it is concerned about any "governmental action that interferes with the willingness of customers to purchase books,"[17] but how do you define willingness? Some individuals may not be willing to buy certain books (e.g., adult erotica) not just because the government might find out about it, but that someone standing in line at the local store might recognize them or a family member might sign for the book after Amazon ships it. Should we put private booths in the checkout lines at bookstores so as not to discourage purchases?

Protecting private choice in the absence of privacy requires continual vigilance and effort, but it is something in which society already engages on a regular basis. We see this in efforts to defend individuals against discrimination. For example, in 1948 a famous social psychology experiment was conducted on prejudice.[18] After World War II, a Canadian researcher mailed two reservation requests each to one hundred resorts in Ontario. On one reservation he used the name "Mr. Lockwood," and 93 percent of the resorts offered reservations. On the other reservation, he used the name "Mr. Greenberg" (frequently a Jewish name), and only 36 percent responded favorably. If the logic of the modern privacy debate had been applied to this issue back in those days, regulations would have been issued requiring that reservations be made anonymously lest individuals be chilled from expressing their constitutionally protected right to travel.

Of course, it makes no sense to try to hide the fact that a person is Jewish, anymore than it would for a person of color to wear a bag over his head to hide his race. This is not how society has approached the problem; rather, we focus on ensuring that no one is denied access to education, government services, employment opportunities, or public facilities regardless of personal details like race or ethnicity. When this approach is extended to other forms of personal data in the public domain, we have a model for how freedom can exist in the face of limits to privacy.

A more focused view of privacy and vigorous protection of private choice is our best hope for dealing with the realities of the twenty-first century where information "longs to be free." It recognizes that information willingly turned over becomes part of the public domain. It accepts the fact that although the right to free speech does not provide the right to harass individuals, it protects the exchange of data between companies. It warns us to use caution with statutory approaches to privacy as the ill-conceived *HIPAA* regulatory regime bears witness. It may even be the case that a less distended form of privacy that returns to its roots of shielding a man's castle will instill greater respect and offer more meaning for the concept than we find today.

Furthermore, if a movement toward greater openness with data is paired with secure identification, the public might be more supportive, knowing that thieves won't be using individuals' personal information by posing as them. The number-one consumer complaint to the FTC is probably one of the main reasons that information has such a bad rap; eliminate identity theft, and it will become clear that our efforts shouldn't have been focused on the data, but on the people who use it under false pretenses and on the system that does little to prevent it.

As Chapter 13 will show, forces at work in modern information economies are driving the train of transparency forward at breathtaking speed. Our best hope is not to try to stop the engine, but to make sure it doesn't run over anyone's most important constitutional freedoms. Perhaps the courts will one day come to this recognition and shift their focus from guarding the anonymity of people going door to door to making sure they always have a door on which to knock.

PART IV

Conclusion

CHAPTER 13

THE OPEN SOCIETY OF THE TWENTY-FIRST CENTURY

Overcoming the Crisis in Trust

We began this discussion by suggesting that the open society of the twenty-first century will have the best chance to remain vibrant, robust, and secure if it is willing to lower the walls that separate us and embrace greater transparency. Instead of a world where people walk around cloaked in anonymity and where the desire for privacy shields those engaged in terror and other abhorrent acts, this book has presented a case for a society where technology makes openness the norm, where public actions, including those of our leaders, are exposed to more scrutiny, and where privacy, while still respected, is more circumscribed and narrowly defined.

Although on the surface, this view of American society may appear to be little more than a sacrifice of privacy at the altar of technology, the movement toward greater openness is about something much more fundamental. At its core, it is about restoring the value of trust to its rightful and necessary place in life.

Trust has always been an essential component of a civil society. Whenever people have formed agreements, contracts, and other arrangements for living and working together, there has been a need to know that others will act in accord with what they say.

Francis Fukuyama in *Trust: The Social Virtues and the Creation of Prosperity* describes trust as a fundamental ingredient of prosperity.[1] Cultures with low levels of trust like those in Latin America, for example, find it difficult to evolve from an economy of family-run businesses to one of thriving, large-scale, corporate enterprises needed to sustain a modern economy. In countries where there are higher levels of trust, like the United States or Japan, citizens are more likely to arrange themselves spontaneously into organizations of increasing richness and complexity, ranging from church groups and professional associations to global conglomerates. Trust thus contributes to what social scientists like Fukuyama call social capital, or the level of social connectedness in a society.

Researchers have paid much more attention to the concept of trust over the last few decades and have documented the precipitous fall in its levels in America. In a much publicized article published in 1995 entitled *Bowling Alone: The Collapse and Revival of American Community*, Robert Putnam used historical data to suggest that America's stock of social capital and trust, which had been flourishing throughout most of the twentieth century, began to decline in the 1960s as fewer people joined civic organizations like bowling clubs, unions, and volunteer groups.[2] Putnam points out that national surveys that ask people whether they usually trust or are wary of others show the same level of decline over the last quarter century.

If trust was already in a tenuous state in America, the 9/11 attacks struck a blow as deadly as those that hit the Twin Towers to what remained of it. The confidence that society put into individuals like al-Mihdhar and al-Hamzi by opening its borders and neighborhoods was betrayed the moment the two chose to hijack an airplane and turn it into a missile. The very goal of terrorism, to cause people fear and anxiety, undermines their basic ability to trust. How many people since 9/11 have been guilty of looking at a dark-skinned Arab on an airplane and feeling a fleeting pang of trepidation?

There have always been and always will be times when the public's trust is breached. The fact that there is a need for trust in the first place, that there isn't a guarantee of a desired outcome, by its very nature means that there can be no assurance that a person will uphold his or her end of the agreement. You may expect a car to stop at the intersection you're crossing, but short of taking all cars off the road, there is no way to be certain that you won't end up in a violent car crash. There are simply risks associated with living in society, and we must accept them if we plan to go on living.

This is not to say that we cannot do our best to minimize certain risks. We've entered an age where millions of people share the same roads, airways, fiber-optic lines, energy sources, shopping stores, food supplies, and so on. In such a dense and highly integrated society, a single action can reverberate through this network and have dramatic consequences for its members. We saw an example of this in the summer of 2003 when a few high-voltage transmission lines in Ohio failed, causing massive power failures in the northeastern United States. We see it on a regular basis with the spread of destructive computer viruses causing billions of dollars' worth of damage around the globe. Now imagine a single action caused by a fanatic armed with weapons of mass destruction who desires to deliver a judgment of death on thousands, if not millions, of individuals. This scenario forces us to ask ourselves whether there is a better way to manage our risks. Is there, so to speak, a way to determine whether we can keep a few of the most dangerous drivers off the road and away from the intersections of life?

Greater openness serves a vital function in this endeavor. Through measures like secure identification, surveillance, facial recognition, and IA, society can make improved judgments of trust not based merely on blind faith, but on factual information, such as public behavior and personal history. Whenever actions have a potential social impact, such as buying a weapon, boarding a plane, purchasing chemicals, or conducting bioengineering research, transparency can increase society's success in choosing in whom to place its faith. September 11 showed us that the results of a poor decision of when to trust made in a critical situation can be disastrous.

The notion of finding improved ways to hold people to their word is not new. The drive toward accountability has taken hold in many corners of society, such as in accounting audits for companies, standardized testing for teachers and students, and videotaping of police officers. Mary Graham in *Democracy by Disclosure: The Rise of Technopopulis* describes how the growth in disclosure laws, which force companies to reveal information about their products and practices, has served to reduce health, safety, and environmental risks to the public.[3] Exposing the levels of chemicals released into the environment, additives put into food, medical errors in hospitals, the risks of SUV rollovers, or the amount of lead in paint allows people to make informed decisions about which companies they should trust to sell them products and services.

The power of the Internet has amplified the process of disclosure by putting tremendous amounts of information at the fingertips of the public. For instance, citizens can go to any number of Web sites, type in their zip code, and get an instant report of environmental quality and scorecards on companies that release toxic chemicals in a given area. As Graham points out, when potentially damaging information energizes consumers to take action, companies are motivated to make changes and restore public confidence with greater speed than any government regulator could ever inspire.

Although a movement toward openness and trust has filtered through a great number of institutions and organizations in society, the individual remains a holdout. A view of privacy that embraces anonymity has placed a cloud of obscurity over the average person on the street in the modern world, making it impossible to know whom to trust. The Internet, with its diverse array of cheap, yet powerful, tools for hiding identities only serves to reinforce this trend. According to Michael Froomkin, a legal scholar and avid defender of anonymity,

> Digital anonymity exacerbates the trends that are producing a society of strangers. Strangers are people who lack the mutual and continuous monitoring associated with life in a small town. Another way of putting the same point is that strangers are people about whom one has little or no information; in effect strangers engage each other as if they had complete informational privacy. A society of strangers may be one in which trust may be more difficult.[4]

Froomkin is on the mark when he acknowledges the reverse correlation between privacy and trust. In other words, the more privacy we have, the less likely we are able to trust someone. Knowing something about a person helps you make a reasonable judgment about whether to trust him or her.

Unfortunately, all the doomsaying about privacy obscures the positive benefits that society will accrue when there are fewer strangers among us. Removing the veil of anonymity and revitalizing trust will give the majority of honest and hard-working Americans the opportunity to showcase their positive reputations. As in the village of the past, where people's family names and personal reputations were their strongest currency, so today can a person's identity be an important asset that facilitates their interactions in society.

When there is no question as to people's identities and reputations, "trusted" Americans and foreign visitors will benefit. It makes no sense that someone's grandmother, who has spent a lifetime as an upright and outstanding citizen, should be patted down at an airport. Instead, she should be allowed to stroll right past the security checkpoint and board a plane more expeditiously than ever.

In the same vein, it is just as deleterious to label people as security risks because of their race or ethnicity. When there are supposedly a few hundred al Qaeda members hidden in the United States, it is foolish to harass millions of honorable and patriotic Americans of Arab descent as potential threats at security sites around the country. Immigration has provided a vibrant and diverse melting pot of individuals who contribute ideas and values, stimulate the economy, and supply an overall cultural richness to the United States that would otherwise be lacking. To turn our backs on this history in a reflexive action to close our borders and round up innocent foreigners would be just as tragic as the 9/11 attacks.

A better approach is to use technology that allows us to bypass human bias, prejudice, and cognitive deficits to target real threats in the United States, much like a GPS guided missile hones in on and destroys a building, leaving everything else around it intact. As we've seen, the technologies of openness expose everyone's public actions so that when something suspicious occurs, such as individuals with potential links to terrorists taking flying lessons or applying for permits to drive hazardous waste vehicles, it stands out clearly.

We already see mechanisms for determining trust being employed by the TSA at airports where the CAPPS II system is designed to categorize people based on risk. Airlines can make much more informed decisions on security matters by using substantive information, rather than relying on whether a passenger has the flimsy, insecure piece of paper known as a driver's license.

Even if a national identification card is implemented, IA systems like CAPPS II should still be utilized to calculate measures of risk, an approach that could combat the first-time terrorist problem described in Chapter 5. For instance, the system can be instructed to weigh a number of variables, perhaps deciding that a foreign visitor who has only been in the country for a few months or has an erratic financial record that includes dozens of bank accounts has to go through a more intensive screening.

In contrast to how CAPPS II has been portrayed, the system is not intended to restrict freedoms; in fact, the technology of openness will put most Americans who live and play by the rules in the fast lane to freedom—in this case, by facilitating the boarding of an airplane. And, as was mentioned in Chapter 5, criticisms that some people will be denied the right to fly based on confusion over names on a terrorist watch list only highlight the need for improved identification.

Better decisions about whom to trust promise to impact much more than just the antiterror campaign. People will appreciate knowing that they can place greater confidence in the caregiver watching an elder parent in a nursing home, in the financial planner managing a retirement account, or in the medical provider writing a prescription. Many abuses will be stamped out at short notice, preventing for instance, an organization like the Catholic Church from stealing the lives of so many innocent children in a decades-long sex scandal. Arguing along these lines is privacy expert Fred Cate, who affirms that society has a legitimate need to expose information that privacy seeks to protect:

> What parent would not want to know if her child's babysitter had been convicted for child abuse? Similarly, what storeowner would not want to know whether his physician had a history of malpractice? What man or woman would not want to know if a potential sexual partner had a sexually transmitted disease? What airline would not want to know if its pilots were subject to epileptic seizures? Yet the interest in not disclosing that information is precisely what privacy protects.[5]

Of course, who wouldn't want to have accurate facts about the risks faced in life? Many fear, however, that leaders in power will use information not just to determine trust, but against individuals, whether for political advantage or to restrict civil liberties. Chapter 10 discusses the depth of anxiety over Big Brother potentially misusing the technologies of openness. As this book and others before it have argued, greater openness needs to be a two-way street, where citizens, the media, and public interest groups can monitor, and thus trust, the actions of their leaders, just as they themselves are monitored. History teaches us that government activities undertaken behind closed doors represent a path fraught with temptation for abuse. Transparency can be the roadblock to prevent our leaders from going down this path.

A fear of openness is further stoked by the public's desire for anonymity, a sentiment with roots that extend far into America's past. Americans have long romanticized the ability to retreat from public view. Many years ago the country's vast frontier represented a chance to flee persecution, hide from one's past, or just live a simple, quiet, and self-sustaining life in the freedom of anonymity. A few centuries later, in our buzzing, interconnected, electronic world, this nostalgic view of life still reverberates among members of the public who reject the idea of being unveiled by technology.

Perhaps the public would be much more accepting of transparency if people were confident that the same level of openness would be directed at the officials manning the cameras, maintaining the watch lists, and monitoring the IA programs. The question asked centuries ago about who would watch the watchers, *quis custodiet ipsos custodies*, has never been more relevant than it is today. The good news is that we've seen a general trend over the last few decades toward more openness in government. Unfortunately, on occasion those in power resist this trend and try to swim against the current. We see such a swim team in the administration of President George W. Bush.

Government Secrecy

Everything secret degenerates, even the administration of justice; nothing is safe that does not show how it can bear discussion and publicity.

—Lord Acton in the nineteenth century

An intractable proclivity toward policies of secrecy seems almost an inherent quality of those in power. Even in the world's oldest democracy, the United States, leaders have a flawed and inconsistent history of supporting openness. Over the last century, on numerous occasions the failure to embrace transparency, even with the best intentions, has led to corruption, scandals, missteps, and abuses of power that have rendered the government ineffective at best and a violator of individual rights and freedoms at worst.

Former senator Daniel Patrick Moynihan in his book *Secrecy* provides a survey of government policies over the last century that have encouraged the overclassification of material and the bureaucratic closeting of information, all to deleterious effects.[6] For instance, he argues that had U.S. security agencies revealed the existence of a small, but manageable, group of communist spies in the country, the fire of McCarthyism and the Red Scare wouldn't have had enough timber to burn. Only in an atmosphere of uncertainty could the latent paranoia about a Communist infiltration of the U.S. government rage into a full blaze. He also points to the Bay of Pigs, which preceded the Cuban Missile Crisis, the Pentagon Papers, which preceded Watergate, and the Iran-Contra affair as examples of the disastrous effects of the rampant secrecy that has plagued the U.S. government.

Moynihan's book resulted from a commission that he cochaired to look into the wall of secrecy the U.S. government had constructed over many years. The Commission on Protecting and Reducing Government

Secrecy began its investigation when signed into law by President Bill Clinton on April 30, 1994. Released after almost three years, many interviews with government officials, and numerous visits to agencies like the CIA and FBI, the commission's report was critical of the government's system of classification, which it claimed had spiraled out of control with the overclassification of documents. It recommended that in weighing the need to classify versus the interest of public disclosure, the government should choose to release documents whenever there is significant doubt.

As part of a spirit of openness that anticipated the findings of this commission, President Clinton issued Executive Order 12958, "Classified National Security Information," which he signed on April 17, 1995. The order initiated one of the largest declassification efforts in modern history, releasing millions of formerly classified documents.

With the election of George W. Bush to the presidency, it seemed like the executive branch made a U-turn toward the closed-door policies of the past. A propensity for secrecy in the administration was even apparent prior to 9/11, before concern over national security reached a new peak. According to the *New York Times*, as of the end of September 2001, the 260,978 classified documents in the executive branch represented an increase of 18 percent over the previous year.[7]

During the early stages of the administration, in a portent of things to come, Vice President Dick Cheney defiantly refused to provide information about the energy executives he met with as part of an energy task force he headed. Concerned about the undue influence of the energy industry in shaping the administration's policies, the GAO took the vice president to court to force the release of the documents, suing the executive branch for the first time. Cheney, in an attempt to stave off attacks, claimed he stood on a matter of principle and was protecting the president's prerogative to get confidential advice.

Since 9/11, numerous actions supporting secrecy in the name of fighting terrorism have evidenced the administration's antipathy toward openness. In June of 2003 when the congressional joint inquiry into 9/11 was released, the administration kept twenty-eight pages supposedly dealing with Saudi Arabia's involvement in the attacks classified.

Another clear example of the administration's desire for secrecy occurred in October 2001 when Attorney General John Ashcroft issued a policy statement advising agencies to withhold information from FOIA requests if there is a "sound legal basis," adding that the "Department of Justice will defend your decisions."[8] The lowering of the standard from the previous administration that encouraged disclosure unless it was "reasonably foreseeable that disclosure would be harmful" has led to a significant increase in the number of denied FOIA requests.

Ashcroft's changing of this standard is not an isolated incident. In a separate measure he attempted to close formerly public immigration hearings, arguing that al Qaeda could obtain a list of detainees to gain insights into how the terrorist hunt was going. Names and the details of arrests made by the INS were also kept private.

Keeping information from terrorists was also the explanation given for the cleansing of many government agency Web sites. Agencies, including the EPA, have pulled information off their Internet sites that provided structural layouts, locations and risk factors of nuclear plants, chemical factories, dams, reservoirs, and pipelines. This has flown in the face of so-called right-to-know and duty-to-warn laws that were created to keep communities aware of the potential health and safety risks of living near these facilities.

In another example of the restriction on information, the administration withdrew sixty-six hundred scientific documents on chemical and germ weapons from public access.[9] Citing risks that the information would fall into the hands of terrorists, the administration pulled some declassified documents that went back to the 1940s and 1950s. In addition, the administration has asked the American Society of Microbiology, the world's largest group of germ professionals, to censor what it publishes in many of its journals, including *The Journal of Bacteriology* and *The Journal of Virology*.

Some of the administration's arguments about the need to protect public safety might be more credible if there wasn't such a noticeable culture of secrecy among officials. In February 2003, it was revealed that the Justice Department had been quietly working on proposed legislation to expand the powers in the *USA PATRIOT Act* without the help of the law-making branch of government, the U.S. Congress. Labeled by some the "Patriot Act II," the eighty-page document was leaked to the Center for Public Integrity, causing uproar among members of Congress. Many advocates who have been fighting for a chief privacy officer in every federal agency have gotten it all wrong; it is a chief openness officer that is needed.

Francis Bacon, the philosopher and champion of science, remarked at the close of the sixteenth century, "Knowledge itself is power." When government has exclusive control of information, it assumes power over others, particularly over those who would hold it accountable. The system of checks and balances in the United States depends on the free flow of information to allow, for example, Congress to oversee the actions of the executive branch. Secrecy also undermines the essential ingredient of liberal democracy, which is the active participation and engagement of the populace. Restricting information subverts this process and enables those in power to avoid having to answer for their actions. James Madison recognized the critical role of information in a democracy: "Knowledge will forever govern ignorance, and a people who mean to be their own governors, must arm themselves with the power knowledge gives. A popular government without popular information or the means of acquiring it, is but a prologue to a farce or a tragedy or perhaps both."[10]

Open Government

Injecting greater transparency into government isn't as simple as making more documents available or responding to additional FOIA requests. The system that classifies documents and gives clearances to employees is in place for very good reasons. Information like specific designs for weapons, the names of undercover intelligence operatives, or U.S. tactics and operations, if released, could give our adversaries an advantage and make the world an immediately more dangerous place. For instance, leaks to the media revealed that the United States was tracking calls made overseas by al Qaeda members from prepaid phone cards, disposable cell phones, phone booths, and Internet phone services. These details likely tipped off al Qaeda to the dangers of using communication equipment, thereby making it much harder for the United States to combat this adversary.

Nevertheless, that specific items could lead to harm if released does not excuse the continual, mindless, and wholesale classification of information, a tendency found too often in government administrations.

One of the most strident criticisms of a government mentality of secrecy was put forth by the 1970 report of the Defense Science Board Task Force on Secrecy.[11] It states that "more might be gained than lost if our nation were to adopt—unilaterally, if necessary—a policy of complete openness in all areas of information."

Unilateral declassification would be a first step toward encouraging other countries to become more transparent. An analogy is to the politics of global trade. When the United States cuts subsidies as a good faith gesture, it encourages the rest of the world to take similar steps.

This approach also recognizes that most information is eventually going to work its way out into the daylight of public knowledge. Moynihan points out that even back in the late 1940s, scientists like Hans Bethe, who worked on the atomic bomb, knew that the progress of science meant that Soviets would soon be able to develop their own nuclear weapon. He says, "In 1945 Bethe had estimated that the Soviets would be able to build their own bomb in five years; thanks to information provided by their agents, they did this in four. That was the edge that espionage gave them: a year's worth, no more."[12]

In an age where the government depends on private companies and public universities around the world to carry out its research, it's virtually impossible for the United States to have a monopoly on information.

The Defense Science Board Task Force on Secrecy recognized this point over thirty years ago, writing "it is unlikely that classified information will remain secure for periods as long as five years, and it is more reasonable to assume that it will become known by others in periods as short as one year through independent discovery, clandestine disclosure or other means."[13]

Sometimes, even when there is an abundance of available information, an enemy can do little with it. During the 2003 Iraq War, the U.S. press exhaustively detailed war plans of the U.S. military down to movement of the troops. Simply by watching the embedded reporters on any news channel, Iraq knew the likely route of the 3rd Infantry Division or the 101st Airborne, which cities they would target, and what strategies they would employ. Even so, Iraq could do little to stop the inexorable march on Baghdad by the much superior British and U.S. forces.

Moynihan suggests that in the modern world, simply having information isn't the advantage it once was:

> The central fact is that we live today in an Information Age. Open sources give us the vast majority of what we need to know in order to make intelligent decisions. Decisions made by people at ease with disagreement and ambiguity and tentativeness. Decisions made by those who understand how to exploit the wealth and diversity of publicly available information, who no longer simply assume that clandestine collection—that is, "stealing secrets,"—equals greater intelligence. Analysis, far more than secrecy, is the key to security.

This suggests the idea presented in Chapter 12 that information isn't the problem; the problem is how it's used. For instance, it's one thing to have the plans for a nuclear weapon; it's another thing to actually build one. A country must still purchase supplies, build a facility, produce the nuclear material, dispose of waste, and then construct, test, and store a weapon. All of these steps take time, expertise, money, and materials. Other hurdles like nonproliferation treaties, International Atomic Energy Agency and United Nations inspection regimes, test-ban treaties, satellite photos, and intelligence sources focus the light of scrutiny on countries to make sure they don't turn detailed information into a finished nuclear bomb. Sometimes simple pressure from the world community can encourage a country to open up its nuclear program, as was the recent case with Libya, which decided it no longer wanted to be a pariah in an increasingly interdependent global economy.

Of course, the nuclear scenario is the easy case. It won't be quite as simple to detect countries, let alone individuals, undertaking the development of the GNR technologies described in Chapter 2, which don't require raw materials or large-scale industries. What can the U.S. government do to prevent a scientist hidden away in some dark lab in a far part of the world from manipulating a few genes of the Ebola virus and turning it into an unstoppable killer? Bill Joy's answer would call for a global crackdown on knowledge, forcing all but a few elites to give up research into potentially destructive technologies, which he calls relinquishment.[14]

The battle against Internet privacy is a good case study of how difficult it can be to plug the dike that holds back information. For example, in August 2000, U.S. District Judge Lewis Kaplan ruled in favor of the Motion Picture Association of America (MPAA) in its case against the hacker magazine *2600* for posting the DeCSS program, software that circumvents DVD copyright protection, on the magazine's Web site at www.2600.com. The judge determined that *2600* was guilty of violating the *DMCA*, which expressly prohibits trafficking in circumvention technology. Although much touted by the MPAA, the ruling was a hollow victory, mainly because by that point the genie was out of the bottle and the code had been widely distributed to millions of computer users across the world.

Police crackdowns and government censures won't make the world more secure in the twenty-first century. If anything, this smacks of a much more sinister Big Brother scenario than anything John Ashcroft could imagine does. Rather, the diffusion of knowledge about new technologies will allow the world community to understand the threats it faces. Only by decoding the genome of the Ebola virus, for instance, will we understand how to combat a derivation of it. This kind of knowledge will allow us to build sensors to alert us to the virus's presence, to develop vaccines to limit its effects, and to create forensic genetic databases to track its source.

Preventing further research into this horrific killer, on the other hand, will only force those with ill intentions underground and make it harder to ferret out their secretive operations. Raymond Kurzweil suggests this idea in a response directed to Joy:

> The reality is that the sophistication and power of our defensive technologies and knowledge will grow along with the dangers. When we have gray goo, we will also have blue goo ("police" nanobots that combat the "bad" nanobots). The story of the twenty-first century has not yet been written, so we cannot say with assurance that we will successfully avoid all misuse. But the surest way to prevent the development of the defensive technologies would be to relinquish the pursuit of knowledge in broad areas, which would only drive these efforts underground where they would be dominated by the least reliable practitioners (e.g., the terrorists).

In an age where binary bits of ones and zeros rocket through vast information pipes that crisscross the globe, the U.S. government should shift its focus from the increasingly futile effort of restricting information to the task of convincing other countries to join it in accepting greater transparency. One model in the private sector that could be held up as an example is the Public Library of Science, a nonprofit organization of scientists and physicians dedicated to making scientific literature freely available to the public through the Internet.[15]

Encouraging countries to divulge their secrets may not unveil every individual terrorist concocting a biological potion in his or her basement, but it will go a long way toward distributing the latest scientific knowledge to the world community so it can act in unison against a future menace.

Privacy Advocacy

Resistance to greater openness isn't just found in the halls of America's security agencies. Among the public, privacy advocacy and civil libertarian groups qualify as the most vocal opponents of the trend toward transparency. Although many of their adherents would agree with the need for greater government disclosure, they are determined to resist individual openness at all costs. These are the defenders of extreme privacy, which they see as a sacred, unbounded right that should face no limits.

In an effort to push a hard-core privacy position, many of the staunchest privacy proponents are not above sensationalizing their message to foment anxiety among the public. Consider some of the books published by privacy advocates over the years. Many are what David Brin has called "shrill, paranoiac rants by conspiracy fetishists who see Big Brother lurking around every corner."[16] The titles alone portend a frightening world on the verge of annihilation by the forces of openness. Here are a few from over the years:

- Jerry Rosenberg, *The Death of Privacy* (1969)
- Arthur Miller, *The Assault on Privacy* (1971)
- John Curtis Raines, *Attack on Privacy* (1974)
- Alan Lemond and Ron Fry, *No Place to Hide* (1975)
- Lester Sobel, ed. *War on Privacy* (1976)
- Robert Ellis Smith, *Privacy: How to Protect What's Left of It* (1979)
- William C. Bier, *Privacy: A Vanishing Value* (1980)
- David Linowes, *Privacy in America: Is Your Private Life in the Public Eye?* (1989)
- Jeff Rothfeder, *Privacy for Sale* (1992)
- Deckle MacLean, *Privacy and Its Invasion* (1995)
- Ann Cavoukian and Don Tapscott, *Who Knows? Safeguarding Your Privacy in a Networked World* (1997)
- Whitfield Diffie and Susan Landau, *Privacy on the Line* (1998)
- Charles Sykes, *The End of Privacy* (1999)
- Reg Whitaker, *The End of Privacy* (1999)
- Simson Garfinkel, *DataBase Nation: The Death of Privacy in the 21st Century* (2000)

Since the passage of the *USA PATRIOT Act* in 2001, privacy advocacy groups have pushed the level of rhetoric and emotionalism to new heights. An ad placed by the ACLU in the *New York Times* on February 25, 2003, has the look of a wanted poster from the Wild West and starts off by saying, "Do you know it is now legal for government agents to . . . break into your home when you are away, conduct a search—and keep you from finding out for days, weeks or months that a warrant was ever issued?"[17] Although a government search of your home might be likely if you have Osama bin Laden's number on your cell phone's speed dial, the hyperbole serves to generate an unnecessary level of anxiety in readers. At the bottom of the ACLU's ad is their prescription for anxiety reduction: Readers are instructed to sign up for the ACLU's newsletter and make a financial contribution.

Scare tactics are one thing, but sometimes the drive to warn the public of an impending privacy Armageddon approaches zealotry. With millions of Americans in shock after the grievous attacks on 9/11 and with thousands of families desperate to discover if their loved ones were alive, one privacy advocate, John Perry Barlow, cofounder of the EFF, suggested that citizens should be more concerned about threats posed by their own government. He unconscionably proclaimed on the day of the 9/11 attacks that the United States has "gradually, subtly, invisibly to most of us, become a police state over the last 30 years. This morning's events are roughly equivalent to the Reichstag fire that provided the social opportunity for the Nazi takeover of Germany."[18] Privacy expert Fred Cate has been critical of this kind of tone within privacy circles. He says, "The dominant feature of the current privacy debate is its irrationality. The drivers are emotional."[19]

The dogma of privacy devotees is so ingrained that those who would speak up and challenge it face harsh rebuke. Dorothy Denning, a professor at the Department of Defense Analysis at the Naval Post Graduate School in California, was reviled for supporting the government's position on encryption during the Clipper Chip debate in the early 1990s. *Wired* published an account of her treatment:

> For defending the government's position and having the nerve to insist that she was duty-bound as a citizen to do so, Denning has been reviled. On the Net, she has been the subject of ridicule on countless newsgroups and listservs, enduring vicious—often sexist—personal insults for simply expressing her opinions on matters crypto. She has been booed and hissed in public appearances. Fellow academics and former colleagues whisper that she has abandoned scholarly distance and hopped into bed with the government. When pressed for a memorable example of vilification, Denning herself recalls, with a mix of bemusement and horror, one of the milder epithets: Clipper Chick.[20]

Some within the privacy movement have admitted to regretting the way that Dorothy Denning was treated. One posting to an Internet discussion list said,

> "Demonizing" our opponents, or making them look like dunces (as with the many "I've never heard of Dorothy Denning before" posts), does not help our cause. In fact, it probably weakens our cause, for two reasons.

First, it cuts off dialog with those we disagree with. Second, we tend to underestimate people we have written off as stooges or dunces.[21]

Esther Dyson, an early board member of the EFF and a writer on technological trends, has criticized the self-righteous stance exemplified by the criticisms of Dorothy Denning: "Both we and the people we disagree with are often too shrill. It's so easy for the self-appointed guardians of the digital world to tell their opponents: 'We're right and you're immoral!'"[22]

Others have faced the same treatment as Dorothy Denning. Joseph Eaton, professor emeritus at the University of Pittsburgh, gave up speaking on and researching the topic of national IDs after fierce pressure from critics.[23] After a decade of silence, he is back with a brave reissue of his book on national IDs. In another example, Larry Ellison, the CEO of Oracle, was vilified after 9/11 when he offered Oracle software free to the government to implement a national ID program.

If the government is to wage the war against terrorism successfully without violating important civil liberties, an intelligent and informed debate would greatly enhance the chances of finding a reasonable balance. When individuals sensationalize these issues by giving out Big Brother awards, passing out national ID cards with Larry Ellison's face on them, or posting personal information of John Poindexter (the former head of the TIA program) on the Web and encouraging people to harass him, they do little to further the debate. At a time when the country is searching for a commonsense approach that balances security against privacy concerns, emotionalism only frustrates these efforts.

Public Attitudes toward Openness

There is already some evidence for greater acceptance of openness and transparency and less concern over privacy among the public. For example, when people are asked about privacy in public opinion polls, a high percentage of respondents indicate that they are concerned. Harper and Singleton point out that unprompted surveys tell another story.[24] When people are prompted to answer questions about privacy, they tend to show concern for the issue. But when not prompted and instead asked to list their most important concerns, privacy is not included among the top issues, such as education, health care, social security, and crime.

Behavioral measures are another indicator of how concerned people are about privacy. By looking at whether people act in ways consistent with their attitudes, you can get a sense of how accurate poll numbers really are. For example, consider the Internet, an environment that some have criticized for having poor privacy standards. By some reports, 75 percent of Internet users have provided credit card numbers online, and 52 percent have provided their SSN online.[25]

Overall, it doesn't take much to tempt people to turn over personal information online. A Jupiter report said that 82 percent of respondents would give personal information over the Web to enter a $100 sweepstakes.[26] According to some studies, Internet users do not even look at privacy policies. According to the *New York Times*, only 0.3 percent of users read Yahoo's privacy policy during a one-month period in 2002.[27] Jeffrey Rosen has acknowledged the disconnect between privacy attitudes and behavior among consumers:

> [T]o the frustration of privacy advocates, Americans don't always seem terribly concerned about the possible misuse of click-stream data. Many of us use credit cards for the most intimate on-line purchases. We willingly accept cookies, and we don't take the time to cover our electronic tracks with cumbersome anonymity providers such as Zero Knowledge. For citizens engaged in day-to-day transactions, convenience often outweighs the theoretical possibility that personal information may be disclosed to strangers.[28]

This was affirmed in a report by the polling firm, Gallop: "A recent Gallup survey finds the majority of Internet users pay heavy lip service to concerns about Internet privacy, but at the same time finds most users pay scant attention to the issue"[29] The Pew Internet Project finds, "Online Americans have great concerns about breaches of privacy, while at the same time they do a striking number of intimate and trusting things on the Internet."[30]

As a result of one too many dire predictions about privacy abuses, the public may be beginning to take some of the concerns about privacy with a grain of salt. As Solveig Singleton put it, "the public's concern for privacy is like the River Platte, a mile wide but only an inch deep."[31]

Forces of Openness

Although some of the public may cautiously accept more openness, there may be little people can do to resist dynamics at work that are inexorably pushing us forward into a world where transparency is the norm. One of the forces underlying this relentless march toward openness is based on society's unquenchable thirst for the fuel of the modern economy—not oil, but information.

Transparency is really about knowledge, about closing the gaps between what society knows and what it wants to know in an effort to improve efficiency, make judgments of trust, enhance the quality of life, and further progress. The best example of this is the economy in which powerful incentives exist to uncover information.

Economists like Nobel Prize–winner Ronald Coase tell us that in a market economy where buyers and sellers attempt to connect, transaction costs often get in the way. For instance, when a collector in the United States wants to buy a rare antique vase, he has to spend time and effort to search out a dealer, evaluate the product, and so on. Now, with the Internet at his fingerprints, this person can go online at Ebay and in a few minutes find the desired vase. The gap that once existed between the buyer and seller is now much more transparent, enabling them to find one another in what Bill Gates calls "friction-free capitalism" in his book, *The Road Ahead*.

The desire for information that helps grease economic transactions can be found at work in many elements of our society. Individuals demand details on a company's accounting practices before investing part of their 401K plans in order to avoid an Enron-like experience. Residents want the facts about a company's emissions before deciding whether it is safe for them to live in close proximity to a factory. Voters want disclosures on campaign funding so they can evaluate the likelihood that an elected official has their best interests at heart. Information fills in these gaps so that members of the public can make the best choices for their lives, in much the same way the Wall Street brokers try to make wise investment decisions.

One of the emerging trends in the open society of the twenty-first century, the tracking of products and people, is a good example of how the thirst for information is driving the world toward greater transparency. Whether it is parents using wireless tags to follow their children's movements or adults using sensors to monitor their elderly parents' actions, manufacturers building radio chips into their products to help manage their inventory or farmers using GPS to track herds of cattle or sheep, we see significant gains in efficiency from shining light on information that was once dark and murky. All of these activities were carried out by humans before the arrival of technology; with technology that facilitates the transparency of information, they can be done much more effectively.

As we saw in Chapter 6, with technologies like GPS, RFID, and facial recognition, it won't be long before tracking systems are cheap and powerful enough to allow virtually any object, including humans, to be tracked. The arrival of "sentient computing," the next great technological advance where ubiquitous computer systems embedded in the environment recognize and respond intuitively to individuals' desires, for instance by notifying them when a friend is in the area, recommending a meal at a nearby restaurant, or turning on a favorite television program when they arrive home, will necessitate systems of identification and tracking. Although this might seem scary to some who are focused on the supposed sinister intentions of Big Brother or corporate data invaders, they fail to see that the actions of millions of individuals seeking information to improve the quality of their lives drives us toward a more open and transparent society.

A Technology Cornucopia

The continued explosion in technology will power our interminable desire for information. If advances in technology appear unprecedented, Raymond Kurzweil, an inventor and futurist, says hold on because we haven't seen anything yet. He frequently makes the point that the technology revolution is only just getting started and that exponential, rather than linear, rates of growth will make the future a technological cornucopia. To help people grasp the power of exponential increase, Kurzweil tells the following story:

> To appreciate the implications of this (or any) geometric trend, it is useful to recall the legend of the inventor of chess and his patron, the emperor of China. The emperor had so fallen in love with his new game that he offered the inventor a reward of anything he wanted in the kingdom. "Just one grain of rice on the first square, Your Majesty." "Just one grain of rice?" "Yes, Your Majesty, just one grain of rice on the first square, and two grains of rice on the second square." "That's it—one and two grains of rice?" "Well, okay, and four grains on the third square, and so on." The emperor immediately granted the inventor's seemingly humble request. One version of the story has the emperor going bankrupt because the doubling of grains of rice for each square ultimately equaled 18 million trillion grains of rice. At ten grains of rice per square inch, this requires rice fields covering twice the surface area of the Earth, oceans included. The other version of the story has the inventor losing his head. It's not yet clear which outcome we're headed for. But there is

one thing we should note: It was fairly uneventful as the emperor and the inventor went through the first half of the chessboard. After thirty-two squares, the emperor had given the inventor about 4 billion grains of rice. That's a reasonable quantity—about one large field's worth—and the emperor did start to take notice. But the emperor could still remain an emperor. And the inventor could still retain his head. It was as they headed into the second half of the chessboard that at least one of them got into trouble.[32]

Kurzweil's story demonstrates how exponential trends can lead to tremendously rapid growth, something that Intel founder Gordon Moore envisioned when he predicted almost forty years ago that the number of transistors on a computer chip would double every year. Whereas a chip held a few thousand transistors a few decades ago, today the number has reached two hundred million with no signs of stopping.

The idea of exponential trends like Moore's Law can be applied to more than just computer chips. Kurzweil shows how DNA sequencing, communications technology, computer storage, decreases in the size of mechanical devices, and other technologically related processes are occurring with exponential speed.[33] These advances should impact all sorts of products in the market, making them cheaper, more powerful, and more efficient. The tools of transparency, being dependent on technology, will be particularly affected.

As more countries advance socially, economically, and technologically and, as a result, contribute more educated people to the global economy, the growth curves that Kurzweil describes will continue to shoot upward. As Kurzweil's story demonstrates, the pace of growth starts off somewhat slowly, only to explode to unimaginable levels. Many believe that this point of rapid acceleration may arrive in the next decade or two as a networked world community becomes better able to leverage the latest scientific research and technological advances and as global prosperity, education, and technology increase the world's pool of educated individuals. If one considers the processing power of the human brain, tapping into the minds of poor peasant children in populous and underdeveloped countries like China and India will be like plugging hundreds of millions of supercomputers into a global Internet.

Accountability in the Global Village

As surveillance begins to make the world more transparent, life today is likely to resemble again life in the village of the past in some ways. Just as in small, self-sustaining communities where very little went unnoticed and nosy neighbors knew who could be trusted, so it will be in a future where surveillance technology predominates. In some ways, the village is an apt metaphor because as was pointed out earlier, technology is really just an extension of human abilities.

Life in a global village is likely to impact our lives significantly as events normally hidden from the public eye become transparent, forcing people to take more accountability for their actions. Brin, who has been shown to be prescient in a number of his descriptions of the coming surveillance revolution, describes life in this village:

> In the village, it wasn't fear of retribution, per se, that kept you from behaving callously toward your neighbors; it was the sure knowledge that someone would tell your mother, and bring shame to your family. Tomorrow, when any citizen has access to the universal database to come, our "village" will include millions, and nobody's mom will be more than an e-mail away. Soon that fellow who laughed on the freeway as he cut you off, nearly causing a chain collision, may not be able to hide behind a shield of anonymity anymore. The kid who swipes an apple from a shouting fruit vendor can expect to get a call on the wrist phone before he runs more than a block away. Would-be burglars will have to be awfully clever, when cheap video cameras in any home automatically alert the police and then track the fleeing intruders down the street. True a con artist may be able to look up facts about your finances, but that intrusion will be outweighed when you call up her rap sheet while she's just getting started with her irresistible pitch.[34]

We've already seen one of these predictions come true as several states have used vehicle black box evidence and surveillance cameras to convict speeders and hit and run drivers who have caused fatal accidents.

An all-seeing public will make it much harder to get away with even the little peccadilloes of life. In what might be called the impending age of accountability, crime and terrorism will offer a very poor return on investment for those who choose it as a lifestyle. Hackers, tax evaders, parents who don't pay child support or who neglect or abuse their children, sex offenders, illegal immigrants, spammers, and Welfare fraudsters will be unable to hide their actions in the light of transparency, and society will be relieved from paying the social and economic costs to support them. Strengthening this outcome will be the restoration of shame to prominence as it once again becomes an effective public tool to sanction those who violate community standards and to discourage others from doing the same.

Some critics have expressed fear of a future where greater transparency leads to improved accountability. When asked about the proliferation of cameras in schools, daycares, and nursing homes, John Roberts, executive director of the ACLU of Massachusetts said, "When people are constantly watched, morale suffers and it's not the kind of environment you want at the work place or at schools."

Although Roberts suggests that cameras in schools or nursing homes create an atmosphere of distrust, they really create accountability for people who are caring for children or adults. In important positions where trust is placed in people to act in a responsible manner, society should be able to make sure that high standards are met.

Those critical of surveillance in the open society of the twenty-first century perhaps feel anxiety that people will be held responsible for their choices in life. For example, if surveillance becomes the norm in the workplace, employees may actually have to put in a full day's work to claim a paycheck. Privacy advocate David Lyon mentions a story involving the use of surveillance cameras in some British stores. He is critical of the fact that although the cameras were originally installed to catch criminals, they were soon turned onto employees to see if they were handling exchanges and refunds properly and if they were being pleasant enough with customers.[35] Is it really too much to ask for employees to meet the standards of their managers? Shouldn't they be expected to work the eight hours per day that they are paid for? The key is to make sure that surveillance increases accountability for everyone, down to the clerk in the mailroom and up to the chairman in the boardroom.

Sometimes the argument for privacy is used as a justification to avoid accountability. Consider the position of one outspoken privacy advocate, Katherine Albrecht, founder of Consumers against Supermarket Privacy Invasion and Numbering (CASPIAN), an advocacy group devoted to fighting supermarket card programs. Privacy groups like CASPIAN have been so critical of supermarkets that during the mad cow scare of early 2004, some stores were afraid of contacting consumers who may have purchased tainted beef for fear of violating their privacy rights. Has privacy become so sacred that someone's health should be sacrificed for it?

According to Albrecht, health should not be considered when it comes to privacy: "What happens when your insurance company or your employer buys your supermarket? Or your drugstore? Or if a record of your purchases gets subpoenaed in a lawsuit? Is there no one who might be interested in whether you eat healthfully, smoke or take prescription meds?"[36]

Furthermore, under the guise of protecting privacy, Albrecht makes the inescapable conclusion that consumers should be able to hide the fact that they smoke from insurance companies. Insurance companies are in the business of adequately judging risks so that they can properly charge people for the coverage they need. When they have to make judgments about risk using incomplete information on who smokes and who does not, costs increase for everyone.

Richard Posner has written about the economic benefits of greater transparency and the accountability it affords. He argues that according to economic theory, the law has given too many rights to the privacy interests of individuals and not enough to the needs of organizations to have knowledge about those individuals.

> We think it wrong (and inefficient) that a seller in hawking his wares should be permitted to make false or incomplete representations as to their quality. But people "sell" themselves as well as their goods. A person professes high standards of behavior in order to induce others to engage in social or business dealings with him from which he derives an advantage, but at the same time conceals some of the facts that people with whom he deals need in order to form an accurate picture of his character. There are practical reasons for not imposing a general legal duty of full and frank disclosure of one's material personal shortcomings—a duty not to be a hypocrite. But each of us should be allowed to protect ourselves from the disadvantageous transactions by ferreting out concealed facts about other individuals that are material to their implicit or explicit self-representations.[37]

Another interesting side effect of increased accountability in the open society of the twenty-first century is that relationships are likely to be more honest. Unlike the sporadically employed character George on the "Seinfeld" television sitcom, who frequently claimed that he was an architect, there will be less opportunity to dissemble about our lives. What we buy, where we go, how much our home costs, what we do for a living—these things will become public knowledge.

The inability to hide certain facts about our lives is already happening with the advent of powerful Internet search engines such as Google. In fact, googling has become the practice of doing Internet searches on other people, such as a date, to find out information before you meet them. Perhaps if you find out that your blind date doesn't pay their bills or has been divorced several times, you might see the person in a different light. The same might be said if you find out that your date won an award for philanthropy.

One might ask if the availability of more personal information would be the undoing of personal relationships. Some supporters of greater privacy seem to think so. Jeffrey Rosen argues that privacy is an important part of intimacy and that the closer one is in a relationship, the more important intimacy becomes.[38] He believes that selectively revealing details about one's life enables one to have various levels of intimacy, while violations of privacy threaten this ability.

I'd suggest that on the contrary, my ability to be intimate with my wife is in no way lessened if my neighbor knows what books I read, what products I buy, what hobbies I engage in, and what foods I eat. Becoming personal with my wife is not a result of some secret knowledge she has that I enjoy vegetarian meals; rather, it is the result of my choosing to spend time with her, engaging in and enjoying the preferences we share, such as eating healthy. The fact that I choose to do this with her and not my neighbor assures that there is intimacy in our relationship.

This idea of intimacy is reflected in dictionary definitions that don't mention privacy. Intimacy is described as being "marked by very close association, contact, or familiarity" or a "warm friendship developing through long association." In fact, a person might have an easier time meeting people who share similar interests in the open society of the twenty-first century. We see intimations of this from online dating services like Match.com that allow individuals to efficiently find a potential partner or networks like Friendster.com that facilitate the interaction of groups of friends.

Richard Wasserstrom suggests that rather than detracting from relationships, more openness will enhance the bonds between people by making them more honest.

> [I]nterpersonal relationships will in fact be better if there is less of a concern for privacy. After all, forthrightness, honesty, and candor are, for the most part, virtues, while hypocrisy and deceit are not. Yet this emphasis upon the maintenance of a private side to life tends to encourage hypocritical and deceitful ways of behavior. Individuals see themselves as leading dual lives—public ones and private ones. They present one view of themselves to the public—to casual friends, acquaintances, and strangers—and a different view of themselves to themselves and a few intimate associates. This way of living is hypocritical because it is, in essence, a life devoted to camouflaging the real, private self from public scrutiny. It is a dualistic, unintegrated life that renders the individuals who live it need-

lessly vulnerable, shame ridden, and lacking in a clear sense of self. It is to be contrasted with the more open, less guarded life of the person who has so little to fear from disclosures of self because he or she has nothing that requires hiding.[39]

Some fear that as it becomes harder to hide one's past, people's faults and mistakes will follow them like a scarlet letter through their lives. Perhaps in a society where everyone's foibles are exposed, people will learn to be more forgiving and less judgmental. Although it will be easier to single out people who are different, it will be just as easy to identify the people who try to discriminate against and harass them. American society has made tremendous strides in becoming more tolerant and accepting of diversity in the past few decades. I was thinking of this a few weeks ago when I saw two young men kissing on a downtown street corner in Washington, D.C., and people were walking by without blinking an eye. This would have caused quite a stir less than a decade ago, as would the 2003 ordaining of a gay bishop in the Episcopal Church. There is little to think the trend will reverse, especially when more of people's own diversities are revealed.

Will there be any privacy for people in a transparent future? Just as the line is being drawn now, it's likely that the society of the twenty-first century will draw a distinction between public interactions and personal life in the home. The courts have consistently taken the stand that behavior that happens behind the walls of one's home should be considered private. The Supreme Court ruling in *Lawrence v. Texas* (2003), which struck down a Texas law banning sodomy in the confines of one's home, is the latest evidence of this attitude.[40] Supporting the courts will be advances in technology that will make it possible for people to buy products that further cloak their homes in a shield of secrecy, just as today people buy security systems to secure the perimeters of their homes.

On the other hand, as we've seen, the trend with the courts has been to rule that there is little expectation of privacy with activities that happen in public. As surveillance becomes even more pervasive and wearable cameras and recording devices become the norm, it may be that any interaction with another person or company, even if it happens within the walls of your home, will be considered open to the public. For instance, in the case of *Commonwealth v. Rekasie* (2001), the Pennsylvania Supreme Court ruled that law enforcement could record a telephone call made to a person in his home without a determination in advance of probable cause.

Going forward, people will realize how futile it is to try to restrict information that makes its way into the public sphere and, instead, will focus their energies on demanding that citizens use information for the betterment of society. Some people will always seek out the latest technologies, such as quantum encryption and lasers that blind surveillance cameras, in an attempt to protect their privacy in public, but most people will accept the transparency of society and retreat to their homes when they seek private respite. Of course, there will be privacy where there has always been privacy, in one's personal thoughts, at least until science catches up with science fiction. But let's cross that bridge when we come to it.

Notes

Chapter 1

1. Johanna Mcgeary and David Van Biema, "The New Breed of Terrorist: An Exclusive Look at the Lives of the Men behind the Strike. Now Dozens of Their Associates May Be at Large in the U.S. What Will Come Next?" *Time* (September 24, 2001), at www.time.com/time/covers/1101010924/wplot.html (accessed September 2003).

2. Quoted by Joseph W. Eaton, *Card Carrying Americans: Privacy, Security and the National ID Debate* (New Jersey: Rowman & Littlefield, 1987), 111.

3. Alexander Hamilton or James Madison, "The Structure of the Government Must Furnish the Proper Checks and Balances between the Different Departments," From the New York Packet, February 8, 1788. *The Federalist Papers*, No. 51.

4. John Locke, *Two Treatises of Government*, ed. Peter Laslett (Cambridge: Cambridge University Press, 1988), 305.

Chapter 2

1. United States Air Force, "Air Force 2025 Study," 1996, at www.au.af.mil/au/2025 (accessed September 2003).

2. U.S. Commission on National Security, "21st Century Road Map for National Security: Imperative for Change, the Phase III Report of the United States Commission on National Security," February 15, 2001.

3. L. Paul Bremer III, "A New Strategy for the New Face of Terrorism," *National Interest*, no. 65-S (Fall 2001).

4. State Department Publication, "Foreign Terrorist Organizations," released 2001.

5. Samuel P. Huntington, *The Clash of Civilizations and the Remaking of a World Order* (New York: Simon & Schuster, 1996), 218.

6. Al Qaeda training manual online at www.usdoj.gov/ag/trainingmanual.htm (accessed September 2003).

7. L. Paul Bremer III, "A New Strategy for the New Face of Terrorism," *National Interest*, no. 65-S (Fall 2001), 24.

8. Patrick E. Tyler, "Feeling Secure, U.S. Failed to See Determined Enemy," *New York Times*, September 7, 2002.

9. John Arquilla and David Ronfeldt, "Fighting the Network War," *Wired* (December 2001).
10. Arquilla and Ronfeldt, "Fighting the Network War."
11. Howard Rheingold, *Smart Mobs: The Next Social Revolution* (Cambridge: Perseus Publishing, 2002).
12. Interview with U.S. Senator Bob Graham on NBC's "Meet the Press" on July 13, 2003.
13. Bill Gertz, "5,000 in U.S. Suspected of Ties to al Qaeda," *Washington Times*, July 11, 2002.
14. *Thomas Burnett et al. v. Al Baraka Investment and Development Corp., et al.* (U.S.D.C. D.C.).
15. Todd S. Purdum, "Locked Up and Patted Down: A Year of Making U.S. Safer," *New York Times*, September 8, 2002.
16. Frank Gardner, "Al Qaeda 'was making dirty bomb'," *BBC News*, January 13, 2003.
17. Noah Shachtman, "How Bad Can a 'Dirty Bomb' Be?" *Wired News*, June 10, 2002.
18. Shachtman, "How Bad Can a 'Dirty Bomb' Be?"
19. "Bush Defends Shadow Government," *BBC News*, March 2, 2002.
20. Jong-Heon Lee, "North Korea Amasses Chemical Weapons," *United Press International*, September 17, 2002.
21. U.S. Congressional Office of Technology Assessment, "Proliferation of Weapons of Mass Destruction—Assessing the Risks," OTA-ISC-599 (Washington D.C.: U.S. Government Printing Office, August 1993).
22. Robert Taylor, "All Fall Down: One Hundred Kilograms of Anthrax Spores Could Wipe Out an Entire City in One Go. It's Only a Matter of Time before Bioterrorists Strike," *New Scientist* 50, no. 2029 (May 11, 1996).
23. Ehud Sprinzak, "The Great Superterrorism Scare," *Foreign Policy* (Fall 1998).
24. Richard Preston, *The Demon in the Freezer: A True Story* (New York: Random House, 2002).
25. Tucker Carlson, "Pox Americana," *New York Magazine*, October 8, 2001.
26. Johns Hopkins Center for Civilian Biodefense, Center for Strategic and International Studies, "Dark Winter, Bioterrorism Exercise Final Script," Andrews Air Force Base, June 22–23, 2001.
27. Bill Joy, "Why the Future Doesn't Need Us," *Wired Magazine* (April 2000).
28. Ray Kurzweil, "In Response to," at www.kurzweilai.net/meme/frame.html?main=/articles/art0225.html?m percent3D17 (accessed September 2003).
29. Michael Crichton, *Prey* (New York: HarperCollins, 2002).
30. Rachel Nowak, "Disaster in the Making: An Engineered Mouse Virus Leaves Us One Step Away from the Ultimate Bioweapon," *New Scientist* (January 13, 2001).
31. Nowak, "Disaster in the Making."
32. Sylvia Pagan Westphal, "How to Make a Killer Virus," *New Scientist* (July 20, 2002).
33. Robert Roy Britt, "Survival of the Elitist: Bioterrorism May Spur Space Colonies," *Space.com*, October 30, 2001.
34. Glenn E. Schweitzer and Carole C. Dorsch, "Superterrorism: Searching for Long-Term Solutions," *The Futurist* (June–July 1999).

35. Richard Bernstein, "On Path to the U.S. Skies, Plot Leader Met bin Laden," *New York Times,* September 10, 2002.
36. Amy Westfeldt, "Terrorist Attacks on World Trade Center Cost NYC $33 billion to $36 billion, Experts Say," *Associated Press,* November 13, 2002.
37. Niall Ferguson, "War Names: Random War, Remote War, Absolute War, Do-It-Yourself-War—New Weapons and New Enemies Are Making War New, Too," *New York Times Magazine,* December 15, 2002.
38. U.S. Department of Defense, *Proliferation: Threat and Response* (Washington: Government Printing Office, November 1997), iii.
39. Judith Miller, "Germs: Biological Weapons and American's Secret War," *New York Times,* November 18, 2001.
40. Bill Joy, "Act Now to Keep New Technologies Out of Destructive Hands," *New Perspectives Quarterly* (Summer 2000).
41. Bernie Reeves, "Next Move," *Metro Magazine,* September 12, 2001.

Chapter 3

1. *Talley v. California,* 362 U.S. 60 (1960).
2. *McIntyre v. Ohio Elections Commission,* 514 U.S. 334 (1995).
3. Law.com, "U.S. Supreme Court Upholds First Amendment Rights of Jehovah's Witnesses," June 18, 2002.
4. *American Civil Liberties Union of Georgia v. Miller,* 977 F. Supp. 1228 (ND Ga. 1997).
5. *McIntyre v. Ohio Elections Commission,* 514 U.S. (1995).
6. *Buckley v. Valeo,* 424 U.S. (1976).
7. A. Michael Froomkin, "Anonymity in the Balance," draft v.5, p.16, at www.law.miami.edu/~froomkin/articles/balance.pdf (accessed September 2003).
8. Jeremy Shearmur and Daniel B. Klein, "Good Conduct in the Great Society: Adam Smith and the Role of Reputation," at http://lsb.scu.edu/~dklein/papers/goodConduct.html (accessed September 2003).
9. Online at www.zurich.ibm.com/security/idemix (accessed September 2003).
10. David Brin, *The Transparent Society: Will Technology Force Us to Choose between Privacy and Freedom* (New York: Addison-Wesley, 1998).
11. Richard Ericson and Kevin Haggerty, *Policing the Risk Society* (Toronto: University of Toronto Press, 1997).
12. Roger Clarke, "Certainty of Identity: A Fundamental Misconception, and a Fundamental Threat to Security," July 13, 2001, at www.anu.edu.au/people/Roger.Clarke/DV/IdCertainty.html (accessed September 2003).
13. Al Queda training manual, at www.usdoj.gov/ag/trainingmanual.htm (accessed September 2003).
14. CNN.com, "More Anti-terror Arrests in Italy," November 29, 2001.
15. "Morocco Makes More Al Qaeda Arrests," *Reuters,* June 25, 2002.
16. Daniel Williams, "Italy Probing Source of False Documents: Counterfeiters Suspected of Ties to Al Qaeda," *Washington Post,* July 13, 2002.

17. Daniel De Vise, "Terror Hunt May End Privacy at the Library: Patriot Act Opens Records to FBI," *Herald.com,* September 1, 2002.
18. Council on Foreign Relations, "America Still Unprepared—America Still in Danger," October 25, 2002.
19. Norman A. Willox Jr. and Thomas M. Regan, "Identity Fraud: Providing a Solution," *Lexis-Nexis* (March 2002): 7, at www.lexisnexis.com/about/whitepaper/identityfraul.pdf (accessed September 2003).
20. Brin, *The Transparent Society.*
21. Alan Dershowitz, *Why Terrorism Works* (New Haven: Yale University Press, 2002).
22. Gary Marx, "Identity and Anonymity: Some Conceptual Distinctions and Issues for Research," in *Documenting Individual Identity,* eds. J. Caplan and J. Torpey (Princeton, NJ: Princeton University Press, 2001).
23. Jeffrey Rosen, *The Unwanted Gaze: The Destruction of Privacy in America* (New York: Random House, 2000).
24. Federal Trade Commission, "Consumer Privacy Comments Concerning Information Services—P974806," August 21, 1997, at www.ftc.gov/bcp/privacy/wkshp97/comments1/com266.htm (accessed September 2003).
25. John Stuart Mill, "On Liberty" [1859], in *On Liberty and Other Essays,* ed. Elizabeth Rapaport (Indianapolis: Hackett Publishing Co., 1978), 73.

Chapter 4

1. U.S. Public Interest Research Group, "Theft of Identity: The Consumer X-Files," August 1996, 14.
2. FTC press release at www.ftc.gov/opa/2003/01/top10.htm (accessed September 2003).
3. FTC press release at www.ftc.gov/opa/2003/09/idtheft.htm (accessed September 2003).
4. Arkansas Online, "Poll: Online Shoppers Sweat Credit Card Fraud," May 20, 1999.
5. National Consumers Union/Dell/Harris Interactive Poll, "Econsumer Confidence Study," August 2000.
6. Amitai Etzioni, *The Limits of Privacy* (New York: Basic Books, 1999), 103–11.
7. Department of Health and Human Services, "Birth Certificate Fraud," Office of Inspector General, September 2000.
8. Etzioni, *The Limits of Privacy,* 104–105.
9. Etzioni, *The Limits of Privacy,* 105.
10. Etzioni, *The Limits of Privacy,* 105–106.
11. Etzioni, *The Limits of Privacy,* 106.
12. Jon Dougherty, "Battle over Instant Background Check NRA, Critics Say 'Identity Fraud' Not Exclusively Gun Issue," *Worldnetdaily.com,* March 24, 2001.
13. Etzioni, *The Limits of Privacy,* 107–108.
14. Etzioni, *The Limits of Privacy,* 108.

15. Daniel Sieberg, "FBI: Cybercrime Rising Yet Fewer Companies Reporting Incidents," *CNN.com*, April 8, 2002.
16. Mark W. Vigoroso, "Does Crime Pay More on the Web?" *E-Commerce Times*, January 15, 2002.
17. Kevin D. Mitnick and William L. Simon, *The Art of Deception: Controlling the Human Element of Security* (New York: John Wiley & Sons, 2002).
18. Online at www.ummah.net/mhc/hackers.html (accessed September 2003).
19. Barton Gellman, "Cyber-Attacks by Al Qaeda Feared: Terrorists at Threshold of Using Internet as Tool of Bloodshed, Experts Say," *Washington Post*, June 27, 2002.
20. Anick Jesdanun, "Study: Spam Cost U.S. Corporations $8.9 Billion," *Associated Press*, January 3, 2003.
21. Garrett Hardin, "The Tragedy of the Commons," *Science* 162 (1968), 1243–48.
22. Statement by Robert Douglas before the Interagency Public Forum Hosted by the Federal Deposit Insurance Corporation, "Is It Any of Your Business: Consumer Information, Privacy, and the Financial Services Industry," March 23, 2000, at www.fdic.gov/news/conferences/transcript.html (accessed September 2003).
23. Department of Health and Human Services, "Birth Certificate Fraud," Office of Inspector General, September 2000, at http://oig.hhs.gov/oei/reports/oei-07-99-00570.pdf (accessed September 2003).
24. Simson Garfinkel, *Database Nation: The Death of Privacy in the 21st Century* (Sebastopol, CA: O'Reilly and Associates, 2000).
25. Bob Sullivan, "The Darkest Side of ID Theft: When Impostors Are Arrested, Victims Get Criminal Records," *MSNBC.com*, March 9, 2003.
26. David Brin, *The Transparent Society: Will Technology Force Us to Choose between Privacy and Freedom* (New York: Addison-Wesley, 1998).

Chapter 5

1. Pete Yost, "Feds with Fake IDs Get Past Border Guards," *Washington Post*, January 30, 2003.
2. Quoted by Norman A. Willox Jr. and Thomas M. Regan, "Identity Fraud: Providing a Solution," *Lexis Nexis* (March 2002), 8.
3. Sylvia Dennis, "Passwords Offer Limited Protection, Says Study," *Computer Canada* (February 12, 1999), at http://articles.findarticles.com/P/articles/mi_m/cgc/is_6_25/ai_538805 6 (accessed September 2003).
4. Charles Choi, "DNA Extractable from Fingerprints," *UPI Science News*, July 31, 2003.
5. SEARCH, U.S. Bureau of Justice Statistics, and Alan F. Westin, "Public Attitudes toward the Uses of Biometric Identification Technologies by Government and the Private Sector," January 7, 2003, at www.pandab.org/biometricsurvey.html (accessed September 2003).
6. Will Knight, "Finger Print Smart Card Boosts Security," *New Scientist* (January 2, 2002).

7. Shane Ham and Robert D. Atkinson, "Modernizing the State Identification System: An Action Agenda," Progressive Policy Institute whitepaper, February 7, 2002, at www.ppionline/ppi_ci.cfm?knlgAreaID=140&subsecid=290&contentid=250175 (accessed September 2003).

8. Barnaby J. Feder, "Face-Recognition Technology Improves," *New York Times*, March 14, 2003.

9. Ham and Atkinson, "Modernizing the State Identification System."

10. "Man Charged with Aiding Hijackers," *Associated Press*, September 25, 2001.

11. Phyllis Schafly, "ID Card: The Password to the Police State," October 10, 2001, at www.eagleforum.org (accessed September 2003).

12. Deborah Amos, "Critics: Terrorists Manipulate Loopholes in U.S. Gun Laws," National Public Radio (audio), November 15, 2002.

13. Jon Dougherty, "Battle over Instant Background Check: NRA, Critics Say 'Identity Fraud' Not Exclusively Gun Issue, *Worldnetdaily.com*, March 24, 2001.

14. Amitai Etzioni, *The Limits of Privacy* (New York: Basic Books, 1999), 127.

15. Oliver Burkeman, "U.S. City Where You Can Be Guilty Until Proven Innocent," *The Guardian*, August 27, 2002. There is no reason to think that a Homeland ID will engender more police requests for identification. However, if there is a reasonable suspicion (not necessarily probable cause) that a crime has been committed, police will be within their constitutional limits to stop people under Terry Laws, and under a more recent ruling in *Hiibel v. Sixth Judicial District Court of Nevada* (2004), permitted to ask for identification.

16. "Due Process Vanishes in Thin Air," *Wired News*, April 8, 2003.

17. James K. Glassman, "Time for a National ID Card?" *Washington Times*, November 4, 2001.

18. Steve Connor, "Take Everyone's DNA Fingerprint, Says Pioneer," *The Independent*, February 3, 2003.

19. Julia Scheeres, "Fears about DNA Testing Proposal," *Wired News*, March 31, 2003.

20. *Justice Department v. Reporters Committee*, 489 U.S. 749, 780 (1989).

21. Robert Kuttner, "Why a National ID Card," *Washington Post*, September 6, 1993.

22. Glassman, "Time for a National ID Card?"

23. Brian Pendreigh, "Identity Parade," *Scotsman* (May 25, 1995), 17.

Chapter 6

1. Benny Evangelista, "Surveillance Society: Don't Look Now, But You May Find You're Being Watched," *San Francisco Chronicle*, September 9, 2002.

2. May Wong, "Camera Surveillance Business Is Booming," *Associated Press*, July 29, 2001.

3. Dean E. Murphy, "As Security Cameras Sprout, Someone's Always Watching," *New York Times*, September 27, 2002.

4. Evangelista, "Surveillance Society."

5. Ivan Amato, "Big Brother Logs On," *Technology Review* (September 2001).

6. Gregory T. Huang, "Monitoring Mom: As Population Matures, So Do Assisted-Living Technologies," *Technology Review* (July/August 2003).

7. Roxanne Nelson, RN, "Hidden Cameras in Hospitals Can Uncover Child Abuse: Surveillance Can Also Clear Parents of Wrongdoing," *WebMD Medical News*, June 7, 2000.
8. Dean Takahashi, "Security Cameras Are Getting Smart—And Scary," *Wired News*, January 6, 2003.
9. Marcia Biederman, "In Crash Data, Lots to Debate," *New York Times*, October 23, 2002.
10. Steve Irsay, "Private Surveillance Cameras Play Increasing Role As Investigation Tool," *Court TV*, October 18, 2002.
11. Cited by Spencer S. Hsu, "D.C. Forms Network of Surveillance," *Washington Post*, February 17, 2002.
12. Irsay, "Private Surveillance Cameras."
13. Richard Ericson and Kevin Haggerty, *Policing the Risk Society* (Toronto: University of Toronto Press, 1997), 224.
14. Al Webb, "'Spy' Cameras vs. Villains in Britain," *United Press International*, March 8, 2002.
15. Martin Kasindorf, "High Tech Investigative Tools Get a Push," *USA Today*, October 17, 2002.
16. Hsu, "D.C. Forms Network of Surveillance."
17. S. G. Davies, "The Case Against: CCTV Should Not Be Introduced," *International Journal of Risk, Security and Crime Prevention* 1, no. 4 (1998), 327–31.
18. Webb, "'Spy' Cameras."
19. Todd S. Purdum, "Locked Up and Patted Down: A Year of Making U.S. Safer," *New York Times*, September 8, 2002.
20. Online at www.nice.com (accessed September 2003).
21. Online at www.cernium.com/products (accessed September 2003).
22. Joseph W. Eaton, *Card Carrying Americans: Privacy, Security and the National ID Debate* (New Jersey: Rowman & Littlefield, 1987).
23. Burton Gellman, "In U.S., Terrorism's Peril Undiminished," *Washington Post*, December 24, 2002.
24. Julia Scheeres, "Nuke Reactor: Show Me Your Face," *Wired News*, August 9, 2002.
25. ACLU, "ACLU Calls for Public Hearings on Tampa's 'Snooper Bowl'," Press Release, February 1, 2001, at http://archive.aclu.org/news/2001/n020101a.html (accessed September 2003).
26. "Report: Carnivore Respects Privacy," *USA Today*, November 22, 2000.
27. Jeffrey H. Smith and Elizabeth L. Howe, "Federal Legal Constraints on Electronic Surveillance," in *Protecting America's Freedom* In Th*e Information Age*, report published by the Markle Foundation, October 2002, at www.markletaskforce.org (accessed September 2003).
28. Orin S. Kerr, "Internet Surveillance Law after the U.S.A. Patriot Act: The Big Brother That Isn't," *Northwestern University Law Review* 97, no. 2 (2003).
29. *California v. Greenwood*, 486 U.S. 35 (1988).
30. *United States v. Karo*, 468 U.S. 705 (1984).
31. Elliot Zaret, "Security Versus Privacy," *DCLaw.org*, October 2002.
32. *Florida v. Riley*, 488 U.S. 445 (1989).
33. Zaret, "Security Versus Privacy."
34. Zaret, "Security Versus Privacy."
35. Dana Hawkins, "Cheap Video Cameras Are Monitoring Our Every Move," *USNews.com*, January 17, 2000.

36. C. Norris and G. Armstrong, "The Unforgiving Eye: CCTV Surveillance in Public Space," Centre for Criminology and Criminal Justice, Hull University, Hull, U.K., 1997).
37. Evangelista, "Surveillance Society."
38. Dana Hawkins, "Cheap Video Cameras Are Monitoring Our Every Move," *USNews.com*, January 17, 2000.
39. Simson Garfinkel, *Database Nation: The Death of Privacy in the 21st Century* (Sebastopol, CA: O'Reilly and Associates, 2000).
40. Jeffrey Rosen, *The Unwanted Gaze: The Destruction of Privacy in America* (New York: Random House, 2000).
41. Amato, "Big Brother Logs On."
42. Murphy, "As Security Cameras Sprout."
43. John Markoff, "Protesting the Big Brother Lens, Little Brother Turns an Eye Blind," *New York Times*, October 6, 2002.
44. Charles J. Sykes, *The End of Privacy* (New York: St. Martin's Press, 1999), 186.
45. John D. Woodward Jr., "Privacy vs. Security: Electronic Surveillance in the Nation's Capital," testimony before the Subcommittee on the District of Columbia of the Committee on Government Reform, United States House of Representatives, on March 22, 2002, at www.rand.org/publications/ct/ct194/ct194.pdf (accessed September 2003).
46. Beth Stackpole, "RFID Finds Its Place," *Electronic Business*, June 15, 2003.
47. Robin Clewley, "How Osama Cracked FBI's Top 10," *Wired News*, September 27, 2001.

Chapter 7

1. 9/11 Report: Joint Congressional Inquiry, Report of the Joint Inquiry into the Terrorist Attacks of September 11, 2001—By the House Permanent Select Committee on Intelligence and the Senate Select Committee on Intelligence, July 24, 2003, p. 14, at http://news.findlaw.com/hdocs/docs/911rpt/index.html (accessed September 2003).
2. Richard Shelby, "September 11 and the Imperative of Reform in the U.S. Intelligence Community," in 9/11 Report, at http://news.findlaw.com/hdocs/docs/911rpt/index.html (accessed September 2003).
3. Brian Friel, "FBI Agent: Break Down the Intelligence 'Wall'," *GovExec.com*, September 23, 2002.
4. Friel, "FBI Agent."
5. Section 103-3 of the National Security Act, 18 U.S.C. 403-3.
6. Robert M. McNamara Jr., "A Primer on the Changing Role of Law Enforcement and Intelligence in the War on Terrorism," in *Protecting America's Freedom in the Information Age*, report published by the Markle Foundation, October 2002.
7. 9/11 Report, at http://news.findlaw.com/hdocs/docs/911rpt/index.html (accessed September 2003).
8. Gregory F. Treverton, "Set Up to Fail," *GovExec.com*, September 1, 2002.

9. Michael Isikoff, "Hiding in Plain Sight: Did a Muslim Professor Use Activism as a Cloak for Terror?" *Newsweek*, March 3, 2003.

10. White House Press Release, "Fact Sheet: Strengthening Intelligence to Better Protect America," January 28, 2003, at www.whitehouse.gov/news/releases/2003/01/20030128-12.html (accessed September 2003).

11. Laura Rozen, "Information Sharing at the FBI," in *Protecting America's Freedom in the Information Age*, report published by the Markle Foundation, October 2002.

12. Richard Shelby, "September 11 and the Imperative of Reform in the U.S. Intelligence Community," in 9/11 Report, at http://news.findlaw.com/hdocs/docs/911rpt/index.html (accessed September 2003).

13. Mark Hosenball, Tamara Lipper, Eleanor Clift, Andrew Murr, and Jamie Reno, "The Hijackers We Let Escape," *Newsweek*, June 5, 2002.

14. Markle Foundation, *Protecting America's Freedom in the Information Age*, October 2002.

15. Hosenball et al., "Hijackers."

16. Stewart A. Baker, "The Regulation of Disclosure of Information Held by Private Parties," in *Protecting America's Freedom in the Information Age*, report published by the Markle Foundation, October 2002.

17. Baker, "The Regulation of Disclosure."

18. Eric Braverman and Daniel Ortiz, "Federal Legal Constraints on Profiling and Watch Lists," in *Protecting America's Freedom in the Information Age*, report published by the Markle Foundation, October 2002.

19. Braverman and Ortiz, "Federal Legal Constraints."

20. Paul Rosenzweig, "Proposals for Implementing the Terrorism Information Awareness System," *The Heritage Foundation*, August 7, 2003.

21. William Safire, "You Are a Suspect," *New York Times*, November 14, 2002.

22. Ryan Singel, "Total Info System Totally Touchy," *Wired News*, December 2, 2002.

23. Singel, "Total Info System Totally Touchy."

24. Rosenzweig, "Proposals."

25. Greg Allen, "Prospect of War Raises Fears of Internment," "Morning Edition," National Public Radio, March 11, 2003.

26. See "The September 11 Detainees: A Review of the Treatment of Aliens Held on Immigration Charges in Connection with the Investigation of the September 11 Attacks," U.S. Department of Justice, Office of the Inspector General, June 2003, at www.usdoj.gov/oig/special/03-06 (accessed September 2003).

27. Jennifer Ludden, "N.J. Officials Focus on Frayed Relations with Muslims," National Public Radio, February 21, 2003.

28. Rosenzweig, "Proposals."

Chapter 8

1. Robert Ellis Smith, *Ben Franklin's Web Site: Privacy and Curiosity from Plymouth Rock to the Internet* (Providence, RI: Privacy Journal, 2000), 127.

2. Smith, *Ben Franklin's Web Site*, 134–35.
3. Smith, *Ben Franklin's Web Site*, 287.
4. Simson Garfinkel, *Database Nation: The Death of Privacy in the 21st Century* (Sebastopol, CA: O'Reilly and Associates, 2000).
5. Smith, *Ben Franklin's Web Site*, 290.
6. Smith, *Ben Franklin's Web Site*, 310–11.
7. Vance Packard, *The Naked Society* (New York: Cardinal, 1964).
8. Alan Westin, *Privacy and Freedom* (New York: Atheneum, 1967), 7.
9. *Katz v. United States*, 389 U.S. 347 (1967).
10. *Griswold v. Connecticut*, 85 S.Ct. 1678 (1965).
11. Robert Bork, *Slouching towards Gomorrah: Modern Liberalism and American Decline* (New York: Regan Books, 1996), 103.
12. *Eisenstadt v. Baird*, 405 U.S. 438 (1972).
13. Garfinkel, *Database Nation*, 7.
14. Whitfield Diffie and Martin Hellman, "New Directions in Cryptography," *IEEE Transactions on Information Theory* (November 1976).
15. Reg Whitaker, *The End of Privacy: How Total Surveillance Is Becoming a Reality* (New York: The New Press, 1999), 107.
16. Jeri Clausing, "The Privacy Group That Took On Intel," *New York Times*, February 1, 1999.
17. Rene Laperriere, "The 'Quebec Model' of Data Protection: A Compromise between Laissez-faire and Public Control," in *Visions of Privacy: Policy Choices for the Digital Age*, eds. Colin J. Bennett and Rebecca Grant (Toronto: University of Toronto Press, 1999), 193.
18. R. Gerstein, "Intimacy and Privacy," *Ethics* 89 (1978), 76–81.
19. Phillip Kurland, "The Private I," *The University of Chicago Magazine* (Autumn 1976), 8.
20. Smith, *Ben Franklin's Web Site*, 267.
21. Jeffrey Rosen, *The Unwanted Gaze: The Destruction of Privacy in America* (New York: Random House, 2000), 20.
22. Francis Fukuyama, *Our Posthuman Future: Consequences of the Biotechnology Revolution* (New York: Farrar, 2002), 106–107.
23. United Nations, "Universal Declaration of Human Rights," December 10, 1948, at www.un.org/overview/rights.html (accessed September 2003).
24. Electronic Privacy Information Center and Privacy International, "Privacy and Human Rights 2002: An International Survey of Privacy Laws and Developments," 2002, 17, at www.privacyinternational.org/survey/phr2002 (accessed September 2003).
25. Amitai Etzioni, *The Limits of Privacy* (New York: Basic Books, 1999), 191.
26. *Vernonia School District 47J v. Acton*, 515 U.S. 646 (1995).
27. *Laird v. Tatum*, 408 U.S. 1, 22–23 (1972).
28. Etzioni, *The Limits of Privacy*, 188.

Chapter 9

1. Colin Bennett, "What Government Should Know about Privacy: A Foundation Paper" (paper prepared for the Information Technology Executive Leader-

ship Council's Privacy Conference), June 19, 2001, 2–3, www.gov.on.ca/MBS/english/fip/pub/wgskap.doc (accessed September 2003).

2. David H. Flaherty, *Privacy in Colonial New England* (Charlottesville: University Press of Virginia, 1972), 15.

3. Flaherty, *Privacy in Colonial New England*, 83.

4. Robert Ellis Smith, *Ben Franklin's Web Site: Privacy and Curiosity from Plymouth Rock to the Internet* (Providence: Privacy Journal, 2000), 81.

5. Smith, *Ben Franklin's Web Site*, 50.

6. Smith, *Ben Franklin's Web Site*, 50.

7. Smith, *Ben Franklin's Web Site*, 50.

8. Smith, *Ben Franklin's Web Site*, 108.

9. Alan Westin, *Privacy and Freedom* (New York: Atheneum, 1967), 348.

10. Stewart Kiritz and Ralph H. Moos, "Physiological Effects of Social Environments," *Psychosomatic Medicine* 36, no. 2 (March–April 1974), 96–114.

11. J. H. Earls, "Human Adjustment to an Exotic Environment: The Nuclear Submarine," *Archives of General Psychiatry* 20, no. 1 (1969), 117–23.

12. R. B. Lee, *Dobe Ju/'Hoansi*, 2d ed. (New York: Harcourt Brace College Publishers, 1993).

13. Simson Garfinkel, *Database Nation: The Death of Privacy in the 21st Century* (Sebastopol, CA: O'Reilly and Associates, 2000), 258–59.

14. Adrian Randall, *Before the Luddites* (Cambridge: Cambridge University Press, 2002).

15. Toby Lester, "The Reinvention of Privacy," *The Atlantic Monthly* (March 2001).

16. Maria De La O, *The Industry Standard* editorial review on Amazon.com of *The Code Book: The Science of Secrecy from Ancient Egypt to Quantum Cryptography*, at www.amazon.com/exec/obidos/tg/detail/-/0385495323//qid=1086917863/sr=8-1/r ef=pd_ka_1/002-3647829-1522440?v=glance&s=books&vi=reviews (accessed September 2003).

17. Sarah Stirland, "Mr. McNealy Gets Starry-Eyed," *Red Herring*, September 1, 1999.

Chapter 10

1. Charles J. Sykes, *The End of Privacy* (New York: St. Martin's Press, 1999), 19.

2. Ann Davis, "The Body as Password," *Wired Magazine*, July 1997.

3. Michael T. Kaufman, *Soros: The Life and Times of a Messianic Billionaire* (New York: Random House, 2003).

4. Online at www.soros.org (accessed September 2003).

5. Molly Moore, "Cybermania Takes Iran by Surprise: Youths Swarm Online; Tehran Scrambles to Respond," *Washington Post*, July 4, 2001.

6. Jennifer Lee, "U.S. May Help Chinese Evade Net Censorship," *New York Times*, August 30, 2001.

7. Lee, "U.S. May Help Chinese Evade Net Censorship."

8. Fred H. Cate, *Privacy in the Information Age* (Washington D.C.: The Brookings Institution, 1997), 77.
9. Cate, *Privacy in the Information Age.*
10. Brock N. Meeks, "Congress Targets Privacy Issues, Few Advances Made at Federal Level," *MSNBC.com*, December 8, 2002.
11. Title 50 U.S.C. (a)(3)(A), at www4.law.cornell.edu/uscode/50/1805.html (accessed September 2003).
12. Julia Scheeres, "Feds Doing More Secret Searches," *Wired News*, May 9, 2003.
13. Susan Schmidt, "Lawyers for FBI Faulted in Search Panel Told Legal Staff Misunderstood FISA," *Washington Post*, September 25, 2002.
14. 9/11 Report: Joint Congressional Inquiry, Report of the Joint Inquiry into the Terrorist Attacks of September 11, 2001—By the House Permanent Select Committee on Intelligence and the Senate Select Committee on Intelligence July 24, 2003, at http://news.findlaw.com/hdocs/docs/911rpt/index.html (accessed September 2003).
15. Schmidt, "Lawyers for FBI Faulted."
16. Schmidt, "Lawyers for FBI Faulted."
17. Eric Lichtblau, "U.S. Report Faults the Roundup of Illegal Immigrants after 9/11," *New York Times*, June 2, 2003.
18. Jeffrey Rosen, "Liberty Wins—So Far Bush Runs into Checks and Balances in Demanding New Powers," *Washington Post*, September 15, 2002.
19. Julia Scheeres, "How Changed Laws Changed," *Wired News*, September 11, 2002.
20. Rebecca Smith, *24 Days: How Two Wall Street Journal Reporters Uncovered the Lies that Destroyed Faith in Corporate America* (New York: HarperCollins, 2003).
21. David Brin, *The Transparent Society: Will Technology Force Us to Choose between Privacy and Freedom* (New York: Addison-Wesley, 1998), 135.
22. NPR/Kaiser/Kennedy School Poll, "Attitudes toward Government," National Public Radio, June 2000, at www.npr.org/programs/specials/poll/govt/summary.html (accessed September 2003).

Chapter 11

1. Simson Garfinkel, *Database Nation: The Death of Privacy in the 21st Century* (Sebastopol, CA: O'Reilly and Associates, 2000), 3.
2. Online at www.catmktg.com (accessed September 2003).
3. Richard Ericson and Kevin Haggerty, *Policing the Risk Society* (Toronto: University of Toronto Press, 1997).
4. Paul A. Greenberg, "E-Shoppers Choose Personalization over Privacy," *E-Commerce Times*, January 4, 2000.
5. Reed Stevenson, "Microsoft Says Passport Flaw Exposed User Data," *Associated Press*, May 9, 2003.
6. George A. Akerlof, "The Market for 'Lemons': Quality, Uncertainty and the Market Mechanism," *The Quarterly Journal of Economics* 84, no. 3 (August 1970), 488–500.

7. Walter F. Kitchenman, "U.S. Credit Reporting: Perceived Benefits Outweigh Privacy Concerns," The Tower Group, 1999, at www.towergroup.com/public/ras/default_ras.asp?main=search.asp (accessed September 2003).

8. "Acxiom: Online Marketing Info, a Conscience—and a Hot Stock," *Business Week Online*, March 20, 2000.

9. Sonia Arrison, "Consumer Privacy: A Free Choice Approach," *Pacific Research Institute for Public Policy*, 2001, www.pacificresearch.org/pub/sab/techno/privacy (accessed September 2003).

10. Eugene Volokh, "Freedom of Speech and Information Privacy: The Troubling Implications of a Right to Stop People from Speaking about You," *Stanford Law Review* 52 (2000), 1049.

11. Solveig Singleton, "Privacy as Censorship," Cato Institute Policy Analysis no. 295, June 22, 1998, at www.cato.org/pubs/pas/pa-295es.html (accessed September 2003).

12. Singleton, "Privacy as Censorship," 15.

13. Jeffrey Rosen, *The Unwanted Gaze: The Destruction of Privacy in America* (New York: Random House, 2000).

14. Singleton, "Privacy as Censorship."

15. Singleton, "Privacy as Censorship," 9, quoting Sally Engle Merry, "Rethinking Gossip and Scandal," in *Reputation: Studies in the Voluntary Elicitation of Good Conduct*, ed. Daniel B. Klein (Ann Arbor: University of Michigan Press, 1997), 47 (quoting a study of an Andalusian town).

16. Joshua Quittner, "Invasion of Privacy," *Time Magazine* (August 25, 1997).

17. Virginia Postrel, "No Telling," *Reason* (June 1998).

18. Anne Wells, *Who Owns Information? From Privacy to Public Access* (New York: Alfred P. Knopf, 1995).

19. Thomas Crampton, "Pop Stars Learn to Live With Pirates," *International Herald Tribune*, February 21, 2003.

20. Arrison, "Consumer Privacy."

21. Roger Clarke, "The End of Privacy," at www.anu.edu.au/people/Roger.Clarke/DV/Economist9905-L.html (accessed September 2003).

22. Steven L. Nock, *The Costs of Privacy: Surveillance and Reputation in America* (New York: Gruyter, 1993).

23. Nock, *The Costs of Privacy*, 44–45.

24. Eric Hughes, "A Cypherpunk's Manifesto," at www.eff.org/Privacy/Crypto_misc/cypherpunk.manifesto (accessed September 2003).

Chapter 12

1. Bob Woodward, *Bush at War* (New York: Simon & Schuster, 2002).

2. Simson Garfinkel, *Database Nation: The Death of Privacy in the 21st Century* (Sebastopol, CA: O'Reilly and Associates, 2000), 132.

3. Raju Chebium, "Genome Breakthrough May Require Stronger Privacy Laws," *CNN.com*, June 2000, at www.cnn.com/SPECIALS/2000/genome/story/legal.implications (accessed September 2003).

4. National Conference of State Legislatures, "NCSL Genetics Tables: State Genetics Employment Laws," at www.ncsl.org/programs/health/genetics/clone.htm (accessed September 2003).
5. Martha McNeil Hamilton and Warren Brown, *Black and White and Red All over: The Story of a Friendship* (New York: PublicAffairs Books, 2002).
6. Amitai Etzioni, *The Limits of Privacy* (New York: Basic Books, 1999).
7. Associated Press, "Privacy Rules a Headache for Many: Health Workers Struggle to Comply with HIPAA Regulations," *MSNBC.com*, May 12, 2003.
8. From *The Onion*, quoted in Solveig Singleton, "Privacy as Censorship," *The Cato Institute* (Policy Analysis, No. 295), June 22, 1998.
9. *Helen Remsburg, Administratrix of the Estate of Amy Lynn Boyer v. Docusearch, Inc.*, No. 2002-255 (D.N.H.).
10. *Remsburg v. Docusearch, Inc.*
11. Solveig Singleton, "Privacy as Censorship," *The Cato Institute* (Policy Analysis, No. 295), June 22, 1998.
12. Lauren Weinstein, "Cell-Phone Ban Not a Good Call," *Wired News*, December 9, 2002.
13. Thomson, "The Right to Privacy," 310.
14. Thomson, "The Right to Privacy," 313–14.
15. Etzioni, *The Limits of Privacy*.
16. *Tattered Cover, Inc. v. City of Thornton*, 44 (Col., 2002).
17. *Tattered Cover, Inc. v. City of Thornton*.
18. S. L. Wax, "A Survey of Restrictive Advertising and Discrimination by Summer Resorts in the Province of Ontario," Canadian Jewish Congress, *Information and Comment* 7 (1948), 1–13.

Chapter 13

1. Francis Fukuyama, *Trust: The Social Virtues and the Creation of Prosperity* (New York: Macmillan, 1995).
2. Robert D Putnam, "Bowling Alone: America's Declining Social Capital," *Journal of Democracy* (January 1995), 65–78.
3. Mary Graham, *Democracy by Disclosure: The Rise of Technopopulism*. (Washington, D.C.: The Brookings Institution, 2002).
4. A. Michael Froomkin, "Flood Control on the Information Ocean: Living with Anonymity, Digital Cash, and Distributed Databases," *University of Pittsburgh Journal of Law and Commerce* 395 (1996): www.law.miami.edu/~froomkin/articles/ocean.htm (accessed September 2003).
5. Fred H. Cate, *Privacy in the Information Age* (Washington D.C.: The Brookings Institution, 1997), 29.
6. Daniel Patrick Moynihan, *Secrecy* (New Haven, CT: Yale University Press, 1998).
7. Adam Clymer, "Government Openness at Issue as Bush Holds On to Records," *New York Times*, January 3, 2003.

8. Online at www.usdoj.gov/oip/foiapost/2001foiapost19.htm (accessed September 2003).
9. William J. Broad, "A Nation Challenged: Domestic Security; U.S. Is Tightening Rules on Keeping Scientific Secrets," *New York Times*, February 17, 2002.
10. James Madison, letter to W. T. Barry, August 4, 1822, in *The Writings of James Madison*, ed. G. P. Hunt, vol. IX (New York: G. P. Putnam's Sons, 1910).
11. Office of the Director of Defense Research and Engineering, Report of the Defense Science Board Task Force on Secrecy, Washington, D.C., July 1, 1970, at http://fas.org/sgp/othergov/dsbrep (accessed September 2003).
12. Moynihan, *Secrecy*, 143.
13. Office of the Director of Defense Research and Engineering. Report of the Defense Science Board Task Force on Secrecy, Washington, D.C., July 1, 1970.
14. Bill Joy, "Why the Future Doesn't Need Us," *Wired Magazine*, April 2000.
15. Online at www.plos.org (accessed September 2003).
16. David Brin, *The Transparent Society: Will Technology Force Us to Choose between Privacy and Freedom* (New York: Addison-Wesley,1998), 809.
17. ACLU, "ACLU Targets Attorney General's Insatiable Appetite for New Powers with New Full-Page Ads in *Washington Times* and *New York Times*," Press Release, February 25, 2003.
18. David Brin, "Some Notes about Calamity . . . and Opportunity," *Futurist.com*.
19. Toby Lester, "The Reinvention of Privacy," *The Atlantic Monthly* (March 2001).
20. Steven Levy, "Clipper Chick," *Wired* (September 1996).
21. Ed Carp, "Re: Ad Hominum Attacks [was Re: PC Week Clipper Article]," July 3, 1993, posting on Cypherpunks discussion list.
22. Esther Dyson, *Release 2.0: A Design for Living in the Digital Age* (New York: Broadway Books, 1997), 29.
23. See Amitai Etzioni, *The Limits of Privacy* (New York: Basic Books, 1999).
24. Jim Harper and Solveig Singleton, "With a Grain of Salt: What Consumer Privacy Surveys Don't Tell Us," Competitive Enterprise Institute, June 2001, http://papers.ssrn.com/sol3/papers.cfm?abstract-id = 299930 (accessed September 2003).
25. Harris Interactive/Privacy Leadership Initiative Survey, December 2000, 15, cited by Jim Harper and Solveig Singleton, "With a Grain of Salt: What Consumer Privacy Surveys Don't Tell Us," *Competitive Enterprise Institute*, June 2001.
26. Eric Goldman, "On My Mind: The Privacy Hoax," *Forbes.com*, September 26, 2002.
27. Goldman, "On My Mind."
28. Jeffrey Rosen, *The Unwanted Gaze: The Destruction of Privacy in America* (New York: Random House, 2000), 197.
29. Linda Saad, "Few Web Users Paying Close Attention to Internet Privacy Issue," The Gallup Organization, 2000, www.gallup.com/content/login.aspx?ci = 2284 (accessed September 2003).
30. Susannah Fox, "Trust and Privacy Online: Why Americans Want to Rewrite the Rules," *The Pew Internet and American Life Project*, Washington, D.C., 2000.
31. Solveig Singleton, "Privacy as Censorship," *The Cato Institute* (Policy Analysis, No. 295), June 22, 1998, 2.

32. Raymond Kurzweil, "The Age of Spiritual Machines: Annotated Contents of Chapter One: The Law of Time and Chaos," at www.kurzweilai.net (accessed September 2003).

33. Raymond Kurzweil, "The Law of Accelerating Returns," at www.kurzweilai.net (accessed September 2003).

34. Brin, *The Transparent Society*, 333.

35. David Lyon, *Surveillance Society: Monitoring Everyday Life* (Philadelphia: Open University Press, 2001).

36. Kristin Davis, "Loyalty Has Its Price by What You Really Pay by Knuckling Under to Those Bonus-Card Bullies," *Kiplinger.com*, October 2001.

37. Richard A. Posner, "An Economic Theory of Privacy," in *Philosophical Dimensions of Privacy: An Anthology,* ed. Ferdinand David Schoeman (New York: Cambridge University Press, 1984), 338.

38. Rosen, *The Unwanted Gaze*.

39. Richard A. Wasserstrom, "Privacy: Some Arguments and Assumptions," in *Philosophical Dimensions of Privacy: An Anthology,* ed. Ferdinand David Schoeman (New York: Cambridge University Press, 1984), 331.

40. *Lawrence v. Texas*, 539 U.S. (2003).

Index

Abdallah, Abraham, 40
accountability, 200–206
Akerlof, George, 157
Albrecht, Katherine, 202
Alien and Sedition Acts, 141
American Association of Motor Vehicle Administrators, 60
American Civil Liberties Union, 144, 145, 194–95
Americans with Disabilities Act, 170
America's Most Wanted, 94
Amnesty International, 148
anon.penet.fi, 31
anonymity
  benefits, 27–28
  in crime, 33–34
  history, 186
  on the Internet, 30–33
  legal cases, 28–29
anthrax, 18
Applied Digital Solutions, 77
Arquilla, John, 14
Arrison, Sonia, 158, 164
Ashcroft, John, 66, 188
Ashe, Arthur, 172
atomic bomb, 191
Atta, Mohammed, 63
Australian National University, 22
Aviation Security Act, 58

Baker, James, 68
Baker, Stewart, 106, 107
Barlow, John Perry, 195

bin Laden, Osama, 13–14
biological detection, 82
biometrics
  digital templates, 62
  errors in, 58
  facial recognition, 57
  iris scanning, 57
biotechnology, 19
Boyer, Amy Lynn, 173
Brandeis, Louis D., 118, 150
Branscomb, Anne Wells, 162
Braverman, 108
breath locks on cars, 78
Bremer, Paul, 12, 13
Brin, David, 33, 34, 50, 150, 151, 166, 194
Britton, John, 171
Brown, Warren, 170
Bush administration, 10, 188–89

California Department of Transportation, 81
Calvin, William, 24
CAPPS II, 185
Carnivore, 85
CASPIAN, 202
Cate, Fred, 143, 186, 195
Center for Strategic and International Studies, 20
Central Intelligence Agency (CIA)
  Malaysian intelligence contact, 4
  tracking 9/11 terrorists, 98–99
chemical weapons. *See* weapons of mass destruction

Cheney, Dick, 188
Chernobyl, 17
China, 141, 164
Church of Scientology, 31
Clarke, Roger, 34
classification of documents, 188
Clinton administration, 123
Clinton, Bill, 19, 150, 188
Clipper Chip, 122–23
Coase, Ronald, 198
Code of Fair Information Practice, 122
COINTELLPRO, 99, 119
cold war, 11
Communications Assistance to Law Enforcement Act, 123
Computer Fraud and Abuse Act, 43
Computer Matching Act, 71, 107
cookies, 155
copyrights, 162–64
  DVD, 192
Corrigan, Katie, 56
Council on Foreign Relations, 35
Crichton, Michael, 21
Crime Stoppers, 94
Curry, Brian, 90
cybercrime
  auction fraud, 42–43
  Nimba virus, 43
  phishing, 46
  social engineering, 43
cyberterrorism, 43
Cypherpunks, 165

Dark Winter, 20
Defense Advanced Research Projects Agency, 82
Defense Science Board Task Force on Secrecy, 191
Denning, Dorothy, 195–96
Deparment of Homeland Security, 101
Dershowitz, Alan, 36, 69
Devil's Dichotomy, 35
Diffie, Whitfield, 123
digital certificates, 45
Digital Millennium Copyright Act, 32, 162
digital rights, 163–64

digital signatures, 60
digital video recorders, 155
dirty bomb. *See* weapons of mass destruction
DNA, 22, 58, 70
Docusearch, 173
Dorsch, Carole, 23
Doubleclick, 155–56
Drexler, Eric, 21
Dyson, Esther, 196

EarthCam, 90
Eaton, Joseph, 196
Echelon, 85
Edelstein, Herb, 110–11
Edwards, John, 146
Edwards, Jonathan, 131
Electronic Communications Privacy Act, 122
Electronic Frontier Foundation, 124
Electronic Privacy Information Center, 124, 127
Ellison, Larry, 196
equal protection, 108
Ericson, Richard, 33, 80, 154
Espionage Act, 142
Etzioni, Amitai, 41, 127, 129, 171, 175–76
European Convention on Human Rights, 128
event data recorders, 78

Face It, 83
facial recognition, 83–84, 91–92
Faustian bargain, 6
Federal Airline Administration, 38
Federal Bureau of Investigation (FBI)
  FISA surveillance, 144–46
  informant on 9/11 hijackers, 4
  technology concerns, 102
  Ten Most Wanted list, 94
  tracking 9/11 terrorists, 98–99
  Trilogy project, 102
  wiretaps, 123
Federal Privacy Act, 71, 107, 122
Federalist Papers, 6, 27
Ferguson, Niall, 23

Flaherty, David, 131
flight training, 100
Foreign Intelligence Surveillance Act, 144
Foreign Intelligence Surveillance Court (FISC), 99
Founding Fathers, 6, 127
Fourth Amendment, 85–86
Franklin, Ben, 27
Freedom of Information Act, 142
Froomkin, Michael, 29, 184
Fukuyama, Francis, 126, 181–82

Garfinkel, Simson, 89, 135, 139, 153, 168
Gates, Bill, 198
General Accounting Office, 56, 66
General Crimes Guidelines, 108
Gilani, Shiek, 15
Gingrich, Newt, 19
Glassman, James, 73
Global Positioning System, 77
Google, 46, 73, 203
government
  fear of government power, 112
  inefficiency argument, 72
  openness in, 150–51
  oversight, 142–43
  restraints on government, 112
  secrecy, 188–93
  shadow government, 17
Graham, Mary, 183–84
Gramm-Leach-Bliley Act, 161

Haggerty, Kevin, 33, 80, 154
Hamilton, Martha McNeil, 170
al-Hamzi, Nawaf, 3, 27, 38, 97–98, 103–104
Hanjour, Hani, 48, 65
Hardin, Garrett, 44
Hawking, Stephen, 23
Health Insurance Portability and Accountability Act, 36, 170, 171
Hellman, Martin, 123
Helsingius, John, 31
Henderson, D. A., 22
HEW Commission, 121

Howard University, 70
Howe, Elizabeth, 86
Hughes, Eric, 153
Huntington, Samuel, 13

identification
  approach to, 60–63
  criticism of, 63–74
  and daily life, 36
  forged documents, 34
identity theft
  birth certificates, use of, 47
  driver's license, use of, 48
  Identity Theft and Assumption Deterrence Act, 39
  security issue, 49–50
  SSNs, use of, 46
  statistics on, 40
  terrorist use of, 48–49
  types of, 41–42
information
  cost of regulation, 158
  discrimination with, 170–71
  free flow of, 157
  as gossip, 159–61
information analysis
  criticisms of, 110–14
  examples of, 105–106
  legal limits of, 106–109
  pattern-based analysis, 104–105
  safeguards, 109–10
  subject-based analysis, 103–104
information sharing, 97–98, 100–102
INS Passenger Accelerated Service System (INPASS), 69
Internal Revenue Service, 72
Internet Fraud Complaint Center, 43
Internet personalization, 156–57
Interpol, 64
Iqbal, Asif, 69
Iran, 141
Irguns, 12
Israel, 5

Japanese internment during WWII, 112
Jefferson, Thomas, 134

Jones, Paula, 150
Joy, Bill, 20, 24, 192

Kacynski, Theodore, 172
Kazaa, 162
Kelly, Kevin, 160
Kennedy, John F., 119
Kerr, Orin, 86
Kinko's, 47
Kitchenman, Walter, 157
Kurzweil, Raymond, 21, 193, 199–200

Lee, Derrick Todd, 70
Lester, Toby, 136
Lewis, Bernard, 13
Lincoln, Abraham, 142
Locke, John, 6, 127
Lormel, Dennis, 35
Lott, Trent, 150
Luddites, 135
Lyon, David, 202

Madison, James, 190
Maltby, Lewis, 89
Manchausen's Syndrome by Proxy, 77
The Market for Lemons, 157
Marx, Gary, 36
Massachusetts Institute of Technology (MIT), 93
McNealy, Scott, 137
Megan's Law, 67
Microsoft Passport, 156
al-Mihdhar, Khalid, 3, 27, 38, 48, 97–98, 103–104
military preparedness, 4
Mill, John Stuart, 38
Miller, Judith, 24
Miracle Supply, 88
Mitnick, Kevin, 43
Mohammed, Khalid Sheik, 15
Monster.com, 47
Moore's Law, 200
Moussaoui, Zacarias, 35, 145–46
Moyer, Bill, 66
Moynihan, Daniel, 187–88
Muslims of America, 14

mutually assured destruction, 11

nanotechnology, 21
National Data Center, 119
National Driver Registry, 61
national ID, 56
National Instant Criminal Background Check database (NICS), 66
National Intelligence Council, 12
National Liberation Front, 12
National Public Radio, 66
National Rifle Association, 66
National Security Act, 99
National Targeting Center, 83
Neighborhood Watch, 94
9/11, cost of, 23
1984, 140–41
Nixon administration, 120, 121
North Korea, 18
Northern Ireland, 5
nosy neighbor principle, 167
nuclear detection devices, 82
nuclear proliferation, 192

OnStar, 77
Open Society Institute, 141
open society paradox, 8, 53
openness
   attitudes on, 9
   benefits, 7
   driving forces, 198
   opt-in, 9, 158–59
Ortiz, Daniel, 108
Orwell, George, 140–41

Packard, Vance, 120
Paine, Thomas, 27
Patriot Act II, 189
Pearl, Daniel 15
Perritt, Henry H., Jr., 85
Phoenix memo, 100
Poindexter, John, 196
Posner, Richard, 203
Postrel, Virginia, 161
Preston, Richard, 19
Pretty Good Privacy (PGP), 123

privacy
  books on, 194
  chilling effect, 177
  confusion with more basic rights, 174
  congressional action, 121–22
  corporate collection of data, 154–57
  court cases, 120–21
  definition of, 125
  in early America, 131–33
  history of, 118–20
  as a human right, 126–27
  during the industrial age, 131–35
  and liberty, 127–29
  private choice, 9, 175–78
  psychology of, 134–35
  public attitudes on, 196–97
  rhetoric and scare tactics, 194–95
  technology to protect privacy, 136–37
Privacy Rights Clearinghouse, 40
Progressive Policy Institute, 61, 62
psuedonyms, 32
Public Library of Science, 193
Puritans, 131
Putnam, Robert, 182

al Qaeda, 5, 96
  anonymous communications, 35
  as a corporation, 14
  development of chemical weapons, 18
  forged documents, 34
  as a peer-to-peer network, 14
  Suleman Abu Gheith, 16
  tactics, 11, 14–15
  terrorist camps, 15
  training manual, 13, 34

radio frequency identification (RFID), 92–93
relationships, 203–204
remailers, 32
Ressam, Ahmed, 49
Rheingold, Howard, 15
ricin, 18
Roberts, John, 202
Ronfeldt, David, 14

Roosevelt, Franklin D., 118–19
Rosen, Jeffrey, 37, 87, 90, 126, 149, 160, 197, 204
Rosenzweig, Paul, 109, 111, 114, 149
Rotenberg, Marc, 88, 124

Safire, William, 5, 110
Schaeffer, Rebecca, 172
Schlafy, Phyllis, 65, 139
Schulz, William F., 148
Schweitzer, Glenn, 23
search and seizure, 86
secrecy, 188–89
Selective Service, 37
Severe Acute Respiratory Syndrome (SARS), 20
sexual predators, 67
Shelby, Richard, 98
Singh, Simon, 137
Singleton, Solveig, 159–60, 174, 197
smallpox, 19–20
smart cards, 60
smart mobs, 15
Smith, Adam, 30
Smith, Jeffrey, 86
Smith, Robert Ellis, 118, 125, 133, 134
snipers of D.C., 67
Social Security Act, 118
Social Security Administration, 118
Social Security number (SSN), 71, 119
Soros, George, 140
Soviet Union, 11, 16, 19
spam, 30–31, 44–45
Staples, William, 90
State Department, 13
Steinhausler, Fritz, 17
Steve Jackson Games, 124
Stony Brook University, 22
Student and Exchange Visitor Information System (SEVIS), 65
surveillance
  abuse of, 88–89
  aiding accountability, 201–202
  anxiety over, 89–91
  cameras at stop lights, 78
  CCTV, 79

criticisms of, 88–95
in D.C., 76
fighting crime, 79–80
guarding infrastructure, 81–83
limits of, 85–88
protecting airports, 81
protecting borders, 82
reducing accidents, 79
safety use, 77–78
speed cameras, 78
statistics, 75
at the Super Bowl, 83
in war against terrorism, 144–46
Sykes, Charles, 91

Tamil Tigers, 12
Taxpayer Browsing Protection Act, 72
Terrorism Information and Prevention System (TIPS), 95
Terrorism Information Awareness, 7, 105–106
Terrorist Threat Integration Center, 101
Terry Laws, 68
Thomson, Judith Jarvis, 174
TiVo, 156
tragedy of the commons, 44–45
trust, 181–87
Type I errors, 110–12

Universal Declaration of Human Rights, 127
unmanned aerial vehicles, 82

U.S. Air Force, 12
U.S. VISIT, 65
USA PATRIOT Act, 143–50

Viisage, 62
viruses, 22
Voice of America, 141
Volokh, Eugene, 159

wall of separation, 99–100
Walsh, John, 94
Warren, Samuel D., Jr., 118
Wasserstrom, Richard, 204
watch lists, 64
Watergate, 149–50
Watson, James, 70
weapons of mass destruction
  biological weapons, 18
  chemical weapons, 18
  dirty bomb, 17
  International Atomic Energy Association, 17
  spread of, 16–17
web bugs, 155
Weinstein, Lauren, 174
Weisel, Elie, 24
WELL, 31
Westin, Alan, 58, 120, 134
Woodward, Bob, 168
Woodward, John, 88, 90, 92

Zimmerman, Phillip, 123

About the Author

DENNIS BAILEY is an information technology (IT) consultant whose expertise includes security and privacy issues in the public and private sectors. He is a participant in the Subgroup on Identification for the Markle Foundation's Task Force on National Security in the Information Age.

Previously, he was founder and president of onGiving.com, an IT company that raised funds for nonprofits. Before this experience, Mr. Bailey was an IT consultant at such Fortune 500 companies as Sears and GlaxoSmithKline, where he implemented technology solutions that addressed security and privacy concerns.

His education includes a master's degree in political science from American University and a master's degree from the University of Dayton, Ohio. He lives in Alexandria, Virginia.